WINNING THE SLOW RACE

Achieve More by Moving Smarter, Not Faster

OMAR KANDIL

INTRODUCTION

In the fast-paced business world, there's a common misconception that success is synonymous with speed. We're often bombarded with stories of overnight successes, meteoric rises, and instant gratification. What if I told you that the true path to lasting success is not a sprint but a marathon? What if the secret to building something meaningful lies not in chasing shortcuts but in embracing the long, often winding road that leads to sustainable growth?

Welcome to **"Winning the Slow Race"**—a guide for entrepreneurs, business leaders, and dreamers who understand that real success is about playing the long game. This book is not about quick fixes, instant wins, or getting rich overnight. Instead, it's about cultivating the mindset, strategies, and habits to help you build something that stands the test of time.

As someone who has walked the entrepreneurial path, I've learned firsthand that the journey is filled with highs and lows, moments of triumph, and times of doubt. I've made mistakes, faced setbacks, and experienced the pressure to keep up with the business world's breakneck speed. But through it all, I've discovered that the most successful ventures are carefully nurtured, thoughtfully grown, and patiently developed.

In these pages, I will share the lessons I've learned along the way, emphasizing the importance of perseverance, the power of intentionality, and the value of focusing on what truly matters. This book is a compilation of practical advice, real-world examples, and timeless wisdom designed to help you navigate the challenges of building a business in a world that often values speed over substance.

You'll find lessons dedicated to everything from mastering the art of follow-up to understanding why work-life balance is often a myth to realizing that the most important investment you can make is in yourself. Each lesson is a deep dive into a specific aspect of entrepreneurship, offering insights that I hope will resonate with you, regardless of where you are on your journey.

Let's be clear, this book isn't about playing it safe. It's about taking calculated risks, making bold decisions, and learning from every experience, whether a success or a failure. It's about recognizing that shortcuts offer temporary gains but rarely lead to lasting success. Instead, the true rewards come to those willing to work, stay the course, and trust the process.

So, whether you're just starting, looking to take your business to the next level, or simply seeking a fresh perspective, I invite you to join me in exploring the principles of winning the slow race. Together, we'll cover the strategies to help you build a successful business and ensure that your success is sustainable, fulfilling, and aligned with your values.

Thank you for picking up this book. I'm excited to share this journey with you and to help you discover the power of playing the long game. Let's get started.

ABOUT THE AUTHOR

Omar Kandil is the Co-Founder & CEO of OBCIDO Inc. and Talentdu, a digital marketing expert, advisor, and coach with a wealth of experience spanning multiple industries. With over 100 clients, including Fortune 500 companies, Omar has established himself as a dynamic leader in the marketing and business development space.

Born in Syria and raised in Lebanon, Omar's professional journey began in the hospitality industry, where he gained valuable insights into customer service and business operations at a young age. In 2017, he transitioned into marketing, starting with a role at a Dubai-based startup, where he quickly made a name for himself.

Omar's entrepreneurial spirit took flight in 2020 when he co-founded **OBCIDO Inc.** alongside global entrepreneur Chaker Khazaal. Headquartered in New York, OBCIDO has rapidly become a leading boutique creative agency, working with prestigious clients such as **The New York Times, Duggal Visual Solutions, the Library of Congress, STC, Pinto, Lebanese American University,** and **Academic Partnerships**. Under Omar's visionary leadership, OBCIDO has grown significantly, earning recognition for its innovative approach and exceptional client service.

Driven by a passion for education and skill development, Omar also launched **Talentdu**, an EdTech venture dedicated to equipping the next generation of marketers with practical, real-world skills. Talentdu bridges the gap between theory and application, empowering students through hands-on projects and case studies.

Omar is a highly sought-after marketing advisor known for his ability to help global brands and CEOs optimize their strategies and achieve remarkable results. His expertise and insights are regularly shared through his blogs, video content, and speaking engagements, where he continues to inspire and educate others in the marketing field.

CONTENTS

PART 1: MASTERING THE MINDSET

LESSON 1: THE ENTREPRENEURIAL MYTH

There's a pervasive myth in the entrepreneurial world that starting your own business is the only way to achieve real success. Here's the reality: the path to success doesn't always lead through a startup. In fact, in many cases, a corporate job can offer a more stable, lucrative, and fulfilling career than the uncertain journey of entrepreneurship.

The idea isn't that a startup will automatically make you rich, while a corporate job will keep you average. The real question is: what do you want to do with your life? Do you want to build something from scratch, take risks, face challenges head-on, and potentially succeed in a big way? That's the cycle of an entrepreneur. Or, would you prefer to work for a reputable company, grow within that organization, climb the corporate ladder, and eventually become a C-level executive? Both paths are valid, but they offer very different experiences and outcomes.

Let's break it down by the numbers.

1. THE UNICORN SUCCESS RATE

Only a few startups are destined to become the next Stripe, Airbnb, or OpenAI. Achieving unicorn status—reaching a valuation of $1 billion or more—is incredibly rare. As of 2023, there are only 554 unicorns worldwide. To put that in perspective, a startup has just a 0.00006% chance of becoming a unicorn. On top of that, it takes an average of seven years for a startup to grow into a unicorn, assuming it even makes it that far.

2. THE CORPORATE SUCCESS RATE

Now, let's compare that with the success rate of employees in the corporate world. Just in the Fortune 500 companies—which represent the 500 largest companies in the U.S. based on revenue—there are 4,448 C-suite executives. These positions include CEOs (Chief Executive Officers), COOs (Chief Operating Officers), CFOs (Chief Financial Officers), CIOs (Chief Information Officers), CMOs (Chief Marketing Officers), CLOs (Chief Legal Officers), CCOs (Chief Compliance Officers), and CSOs (Chief Security Officers).

And that's just within the Fortune 500. According to Crunchbase, there are 6,612 companies in the U.S. alone that make more than $1 billion in revenue. Now, consider all the other successful companies outside of this list. Do you still think you have a better chance of becoming a unicorn founder versus a C-suite executive at a billion-dollar company?

3. THE FINANCIAL REALITY

Let's talk numbers. How much does a successful startup founder make per year? According to a 2024 report from San Francisco-based accounting platform Pilot, the average startup founder is making $142,000 annually. While that's a respectable salary, it's a far cry from the earnings of a successful corporate executive. In contrast, the median pay package for a CEO in an S&P 500 company was $12.3 million, according to DigitalDefynd.

4. WHY THIS MATTERS

Why am I sharing these statistics? First, to help you choose your path wisely before diving into entrepreneurship. If money is your only goal, becoming a founder might not be the best route. The financial rewards in the corporate world, especially at the executive level, can far surpass those of most entrepreneurs.

Second, I want to dispel the harmful stereotype that demeans 9-to-5 employees. There's a growing trend, particularly on social media, where influencers glorify the hustle of entrepreneurship while belittling those who choose to work for someone else. This mindset is not only misguided but also disrespectful to the millions of professionals who are building successful careers within established companies.

Third, it's important to recognize that no entrepreneur can succeed alone. Your 9-to-5 team will be the backbone of your business's success. They bring stability, expertise, and dedication—qualities that are essential for any company to thrive.

Finally, don't expect your employees to work as hard as you do. As an entrepreneur, you chose a path of uncertainty, long hours, and high risk. Your employees, on the other hand, chose the stability of a 9-to-5 job. Respect that difference. Their commitment is to perform their role to the best of their ability within the framework of their employment—not to match your level of investment in the business.

5. CHOOSING YOUR PATH

In the end, both entrepreneurship and corporate careers offer different paths to success. What's important is choosing the one that aligns with your goals, your risk tolerance, and your vision for the future. Don't let the noise of social media or the myths of entrepreneurship cloud your judgment. Whether you're building a business from the ground up or climbing the corporate ladder, both routes can lead to a fulfilling and prosperous career.

LESSON 2: WORK ON YOURSELF AS HARD AS YOU WORK ON YOUR BUSINESS

As an entrepreneur, your ultimate goal is to see your business grow, thrive, and succeed. Here's a crucial lesson: that growth won't happen unless you're growing, changing, and evolving as well. Your business is a reflection of you, and if you stagnate, so will your company.

It's easy to fall into the trap of thinking that because you've founded an idea, started a business, landed your first client, or even reached the break-even point, you've got it all figured out. The truth is, as an entrepreneur, you don't know it all—in fact, you probably know very little. The journey of entrepreneurship is one of continuous learning, self-improvement, and adaptation. So don't let arrogance or complacency set in just because you've achieved some initial success.

1. IDENTIFY YOUR WEAKNESSES

As you invest in your business—hiring the right talent, perfecting your product, increasing sales, boosting awareness—you also need to invest in yourself. Just as you conduct a SWOT analysis

(Strengths, Weaknesses, Opportunities, Threats) for your business, you should do the same for yourself. Identify your weaknesses the same way you do for your company.

Do you struggle with time management? Leadership? Communication? Negotiation? These are all areas that require your attention. Sometimes, it takes an outside perspective to help you see where you need to improve. This could be a trusted partner, an employee, a mentor, or even a family member. Let them help you identify your weaknesses before focusing on your strengths, and then work hard on those areas.

2. THE DANGER OF COMPLACENCY

A business owner or entrepreneur who reaches a phase where they believe they no longer need to work on themselves is like a company that stops innovating—it's only a matter of time before it fails. Your business can't grow if you don't grow. A comfort zone is a dangerous place for an entrepreneur, and I've seen many fail not because they didn't have a great idea, but because they neglected the importance of self-growth and development.

You need personal development even more than your employees do. As the leader of your business, you set the tone, the vision, and the direction. If you're not constantly improving, learning new skills, and expanding your knowledge, your business will eventually hit a ceiling.

3. EMBRACE CONTINUOUS LEARNING

There are countless opportunities for growth as an entrepreneur. Learn new skills, read more, acquire new technologies, travel, meet new people, connect, attend events, network, and read books. The more you invest in your own development, the more you have to offer your business.

An entrepreneur who thinks they know everything is often the one who knows the least. Don't fall into that trap. Instead, challenge yourself to grow. Choose a new topic every month that you want to become better at, and work on it with the same intensity that you apply to your business.

4. YOUR GROWTH FUELS YOUR BUSINESS

Remember, you are the person leading your business, and it can only grow if you grow as well. The relationship between personal growth and business success is inseparable. By working on yourself as hard as you work on your job, you're not only setting yourself up for personal success but also laying the foundation for a thriving, sustainable business.

So, take the time to invest in your growth. It's the best investment you can make—not just for yourself, but for the future of your business.

LESSON 3: NO IS A FULL SENTENCE

In the journey of entrepreneurship, one of the most powerful tools at your disposal is the ability to say "no." However this isn't just about a verbal "no." It's about a heartfelt "no," a time "no," an energy "no," and even a focus "no." It's about being selective with your commitments so that you can invest your time and energy in the things that truly matter.

As the saying goes, you have to say "no" to many small things to say "yes" to a few big things. Your time is limited, and so is your energy. If you want to succeed, you need to break free from unproductive habits, set clear boundaries, and not worry about disappointing others. Not everyone is on the same path as you. Some people are simply looking to enjoy their time and may invite you along just for company. You need to choose how you spend your time wisely, and don't feel guilty about it.

1. THE 'TO DON'T' LIST

Just as you have a 'to do' list, you should also have a 'to don't' list— a list of things you don't need to do, things that won't move you forward. This could be something as simple as skipping a party with your high school friends. They might not align with your vision and direction, so even if it disappoints them now, when you become a successful founder, they'll all want to reconnect. It's about making choices that align with your goals, even if they're not always popular.

In business, be equally discerning with your time. If you notice that certain activities are consistently wasting your time without moving you forward, find a substitute. For example, I used to get on tens of sales calls every month. What I noticed was that more than 50% of the time, the people I met with couldn't even afford our services. If I spent 20 hours on 20 sales calls per month, that's 10 hours wasted.

2. THE POWER OF VETTING

To address this, I implemented a vetting questionnaire—a form that gets sent to prospects before the call, requiring them to confirm that their resources align with our services. Some people might not appreciate this upfront filtering, but what you're asking for is entirely reasonable. It's a great way to determine early on who's serious and who isn't.

Another example is the time you spend on onboarding, training, creating SOPs (Standard Operating Procedures), and delivering service demos. In a startup, this kind of repetitive work can be overwhelming. Instead of repeating the same tasks over and over, I found a solution that saved me countless hours. I invested in a $20 Loom subscription—a powerful tool that allows you to share video messages with your team and customers to supercharge productivity.

Now, all of our SOPs, training, and service demos are recorded on Loom and can be shared with unlimited people without the need to repeat myself over and over again. This not only saves time but also ensures consistency in the information being shared.

3. RESPECT YOUR TIME, AND OTHERS WILL TOO

Don't worry about your clients—they'll respect your time if you respect it first. By setting boundaries and being selective with your commitments, you're not only protecting your own energy but also demonstrating to others that your time is valuable. This is key to maintaining your productivity and focus on what truly matters.

In the end, saying "no" isn't about shutting doors; it's about opening the right ones. It's about creating the space you need to focus on the big things—the things that will drive your business forward. So, don't hesitate to say "no" when necessary. Your future self will thank you for it.

LESSON 4: THE POWER OF MENTORSHIP

As an entrepreneur, one of the most valuable assets you can have is a mentor. Don't fall into the trap of thinking you know it all—no matter how much you've achieved, there's always more to learn. That's where mentors come in. The key is understanding that your mentor doesn't need to know everything, and in fact, it's often better to have multiple mentors, each guiding you in a specific area of your life or business.

1. THE IMPORTANCE OF MENTORS

Mentors are crucial because they can offer shortcuts to knowledge and wisdom that would otherwise take you years to acquire. They've walked the path before you, faced the challenges you're encountering, and come out the other side with valuable insights. I also believe that age and life experience matter in certain areas. Just because you're a young founder or CEO doesn't mean you can't learn from someone older with more life experience. Their perspective is invaluable, and no matter how hard you work, some lessons only come with time.

However, it's essential to differentiate between true mentors and those who might waste your time, set unrealistic expectations, or ultimately steer you in the wrong direction. Be selective about who you choose to guide you.

2. FINDING THE RIGHT MENTORS

A common struggle many people face is finding mentors, especially if they aren't naturally surrounded by individuals they can learn from. Don't let this discourage you. Depending on the aspect of your life where you need guidance, there are various ways to find the right mentors:

- **LinkedIn:** Reach out to potential mentors who have achieved success in specific areas. Many accomplished professionals are more than willing to share their knowledge and help others grow. Just ensure you're connecting with those who have genuinely succeeded, not those who are simply good at self-promotion.
- **YouTube:** While you won't have direct interaction, YouTube is a treasure trove of valuable content. Industry professionals often share their experiences and insights on almost any topic you can think of. This is a great way to gain mentorship in a more passive form.
- **Mentor Platforms:** Consider using mentor connecting platforms like Intro, Growthmentor, and MentorCruise. These platforms allow you to book sessions with top global mentors, like Alexis Ohanian, for a fee, or connect with unlimited mentors for a monthly subscription. As the importance of mentorship continues to rise, more platforms are emerging, offering access to mentors across various fields. A bit of research can help you find the best option for your needs.

3. THE EVOLVING ROLE OF MENTORS

As your skills and knowledge evolve, you may find that you no longer need the same mentor you once did. That's okay—mentorship is not static. Your needs will change, and so will your mentors. However, one thing remains constant: you will always need a mentor. Even the most successful founders, CEOs, and business owners have mentors to help them grow and make the right decisions.

Never underestimate the power of mentorship. Whether you're just starting out or have been in the game for years, having the right mentors by your side can make all the difference in your journey. So, seek out those who can guide you, learn from them, and continue to evolve. Your growth as an entrepreneur depends on it.

LESSON 5: SUCCESS WITHOUT THE SPRINT

In a world where everyone's chasing overnight success, I've learned to appreciate the value of slow, organic growth—especially when you're bootstrapping. We didn't just avoid investments or venture capital because we lacked options; we chose to bootstrap because every investment becomes a liability, and we wanted to learn and grow along the way.

When you're bootstrapped, growth takes time. It's about building a solid foundation, maintaining a good reputation, and growing at a pace that's sustainable. Guess what? That slow growth? It works. This is not just for bootstrapped startups but also for anyone who wants to build something that lasts. Fast growth might look good on paper, but bubbles often end up bursting.

Imagine this scenario:

YEAR ONE:

- 10 clients served
- 10-15 employees interviewed

- 30 sales calls made

YEAR TWO:

- 25 clients served
- 30-40 employees interviewed
- 60 sales calls made

YEAR THREE:

- 50 clients served
- 50-60 employees interviewed
- 100+ sales calls made

At first glance, it might seem modest. Here's the thing: with each year, more people started to hear about us. Word of mouth began to do its thing. They say that every good experience gets shared several times through word of mouth, and the result isn't linear as you might think—it's exponential.

Year One: Maybe 100 people knew our name.

Year Two: That number grew to around 500.

Year Three: We were on the radar of 2,500 people.

This isn't just growth; it's exponential growth. The best part? It's sustainable because it's built on real relationships, genuine results, and staying true to our values.

Bootstrapped growth isn't glamorous. It doesn't attract headlines or investors. It does something far more valuable—it builds a brand that lasts. Instead of achieving overnight success, it might take you years, but you end up with a solid foundation.

After four years of being a co-founder and CEO, I'm here to tell you that success isn't about how fast you can grow; it's about how strong your foundation is. That's the power of slow, organic growth when you're bootstrapped.

LESSON 6: BUSINESS OVER EMOTIONS

As an entrepreneur, one of the most valuable skills you can develop is the ability to separate your feelings from your business decisions. It's easy to get caught up in emotions, whether it's excitement, fear, attachment, or guilt. Letting these emotions dictate your actions can hurt your business in the long run.

1. THE DANGER OF ACTING TOO FAST BASED ON EMOTIONS

You've probably heard the advice to "trust your gut" or "follow your heart," but what they don't tell you is that before you do, you should stop and think. While some decisions need to be made quickly, others deserve careful consideration. It's about knowing when to act fast and when to take a step back.

For example, decisions like buying a coffee machine for the office, choosing between Microsoft Teams or Zoom, or picking out colors for your brand don't require extensive deliberation. These decisions, while important, don't have long-term consequences that could significantly impact your business. In these cases, trust your instincts and move quickly.

However, when it comes to decisions that do have a long-term impact—like hiring, firing, selecting services, or making significant sales strategies—it's crucial to take your time. Base these decisions on careful analysis, data, and thoughtful consideration, not just on a gut feeling. Acting too fast on emotions in these scenarios can lead to mistakes that are hard to undo.

2. THE PITFALL OF DELAYING DECISIONS DUE TO EMOTIONS

On the other hand, delaying decisions based on emotions can be just as detrimental. Fear is a powerful emotion that can paralyze even the most seasoned entrepreneurs. One of the most common areas where this occurs is when firing an employee. You might know that someone isn't the right fit for the job, but feelings of attachment, concern for their well-being, or the discomfort of conducting an exit interview can cause you to delay taking action.

However, the longer you wait, the more damage it does to your business.. The wrong employee in the wrong position can affect team morale, productivity, and ultimately, your company's success. Do yourself, the employee, and the rest of your team a favor by making the tough decisions, even when they're emotionally challenging.

3. THE PERSONAL DILEMMA: WHEN BUSINESS AND FRIENDSHIPS COLLIDE

I've faced this challenge firsthand. It's tough, especially when the employee in question is also a friend or when your business partner isn't pulling their weight. Imagine having to address the shortcomings of someone you're close to—someone you care about. It's a difficult situation, but it's one that you must handle without letting emotions cloud your judgment.

Now, I make these decisions without involving my personal feelings. It's not easy, but I've learned that it's necessary. Business is not personal, and sometimes, making a tough decision is the best thing you can do for everyone involved. You're not just protecting your business; you're also giving the other person the opportunity to find a role or environment where they can truly thrive.

4. A BALANCED APPROACH

In the end, successful entrepreneurship is about balance. It's about knowing when to act quickly and when to take your time. It's about recognizing when your emotions are helping you make better decisions and when they're getting in the way. By learning to separate your feelings from your business decisions, you'll be better equipped to lead your company with clarity, confidence, and a focus on long-term success.

Remember, in business, it's essential to make decisions based on what's best for the company—not what feels easiest or most comfortable at the moment. By doing so, you're not only ensuring the success of your business, but you're also setting a standard of professionalism and fairness that will benefit everyone involved.

LESSON 7: WHEN YOUR EMOTIONS ARE HIGH, YOUR WISDOM IS LOW

In business, we're often confronted with powerful emotions—both positive and negative. As an entrepreneur, it's crucial to understand that when your emotions are running high, your ability to make wise decisions can be compromised. Emotions create energy, and while that energy can't be eliminated, it can—and should—be controlled and directed properly.

1. THE EMOTIONAL ROLLERCOASTER OF STARTUPS

Startups are a constant rollercoaster of emotions. One day, you might be celebrating the signing of a major yearly contract that you've worked tirelessly to secure. The next day, you might be blindsided by the resignation of your best employee, leaving you without a plan B. These highs and lows are part and parcel of the entrepreneurial journey, but they can lead to impulsive decisions if not managed carefully.

2. THE IMPORTANCE OF TAKING A STEP BACK

When emotions are running high, whether due to excitement or stress, the best advice is to take a step back. If you're experiencing negative emotions and it feels like everything is collapsing around you, resist the urge to make any rash decisions. Instead, pause and reflect. Understand that as an entrepreneur, you don't have to know it all or make every decision on the spot—especially in times of heightened emotions. Take the time to analyze the situation with a clear mind, seek guidance from mentors, get input from your team, and consider your options before taking action.

Similarly, in moments of high excitement and positive energy, it's just as important to take a step back. You might encounter a new technology, trend, or business idea that seems like the next big thing, and you may feel an overwhelming urge to invest all your time, energy, and resources into it. Again, caution is key. Don't get too excited and put all your eggs in one basket. Take the time to think it through, evaluate the potential risks and rewards, and approach it with a balanced mindset.

3. THE DANGER OF ACTING IN THE HEAT OF THE MOMENT

How many times have we reacted in the heat of the moment, only to regret it later? We eventually cool down, but by then, the damage is done. Words and actions can never be taken back, and they can lead to lasting relational damage. Whether it's with family, friends, or coworkers, broken relationships can be difficult to mend, and even when they can be repaired, it takes time and effort.

This is why it's so important to maintain self-control, especially when emotions are high. Lack of self-control can be dangerous, leading to decisions that might satisfy an immediate emotional need but have long-term negative consequences. This brings us back to a concept we've already discussed: trends and investing in them. Just as with trends, emotional decisions should be approached with caution and thoughtful consideration.

4. THE SUBCONSCIOUS INFLUENCE ON DECISION-MAKING

According to research by Harvard Business School professor emeritus Gerald Zaltman, 95% of our purchase decisions are directed by subconscious mental processes. We often believe our decisions are based on logic, but in reality, they're driven by subconscious emotions and only later rationalized with logic. This is a sobering thought, especially when you consider the potential impact of making decisions under the influence of high emotions.

Understanding this subconscious influence can help you recognize when your emotions might be driving your decisions. It's a reminder to take that crucial step back, reassess your situation, and make decisions from a place of clarity rather than impulse.

5. THE ART OF CONTROLLED DECISION-MAKING

In the end, the lesson here is about mastering the art of controlled decision-making. When your emotions are high, remember that your wisdom may be low. By taking a step back, seeking input from others, and giving yourself time to think, you can make decisions that are not only wise but also in the best long-term interest of your business.

In business, as in life, it's not about eliminating emotions—it's about learning to manage them effectively. By doing so, you can navigate the ups and downs of entrepreneurship with a steady hand, ensuring that your decisions are guided by wisdom, not by the whims of the moment.

PART 2: STRATEGIC VISION AND LEADERSHIP

LESSON 8: HIRE FAST, PROMOTE FASTER, FIRE FASTEST

When you're running a startup, recruitment isn't just important—it's crucial. The right team can propel your business forward, while the wrong hires can drag you down. As a founder or CEO, you've got a hundred things on your plate, and recruitment is just one of them. However, here's the reality: waiting too long to make hiring decisions can negatively impact your business.

Gary Vaynerchuk famously said, "Hire fast, promote faster, fire fastest," and I've found this to be incredibly accurate and effective advice for startups. Here's why.

1. HIRE FAST

In a startup, you don't have the luxury of a large HR department to handle the vetting, interview exams, and follow-ups. You're not a massive corporation with a team of ten HR professionals. You're a lean operation and every decision counts. That's why, when you see the right talent—or think you've found the right talent—grasp the opportunity and hire fast.

I know it takes time, and I know that 90% of applicants might end up being a lost investment. However, spending too much time in the hiring process can be a costly mistake. Even with all the right calculations, you can still miss during the interview. Don't let the fear of making the wrong choice paralyze you. Make the decision, and move forward.

2. PROMOTE FASTER

One of the biggest misconceptions I've seen in startups is the idea that you should wait to promote someone, even if they deserve it. There's this outdated notion that an employee needs to wait a year to prove their capacity, as if the passage of time alone validates their abilities. I don't follow this misconception. In fact, I've promoted people four or five months into the business when they've earned it.

The catch, however, is that a promotion should always come with a development plan. Don't just promote a content writer to a marketing director without a clear path for growth. Even if your hierarchy is flat and you only have ten employees, ensure that each promotion is part of a broader plan to develop and sustain your company's talent. Quick promotions can be a powerful motivator, but they need to be backed by a strategy that ensures long-term success.

3. FIRE FASTEST

This is where many startup founders struggle. Firing someone is challenging, especially the first few times. It doesn't feel good, and it requires guts. However, if someone is clearly a wrong fit, you need to act quickly. Keeping them around is like investing in a business you know will keep failing. It's a drain on your resources, your time, and your team's morale.

The best thing you can do for your company, your other employees, and even the person who's not working out is to let them go. It's tough, but it's necessary. Like any other skill, the more you do it, the more you'll get used to it.

SETTING EXPECTATIONS

From the moment you bring someone on board, make it clear that you hire fast, promote faster, and fire fastest. Let them know upfront what to expect. This is a startup, and there's no time to waste—every minute and every dollar counts. By setting these expectations early, you create a culture of high performance, where everyone knows that their contributions are valued, but also that the standards are high.

LESSON 9: EMPOWER WITHOUT HESITATION

In 2024, securing skilled workers is more complex and challenging than ever before. The evolution of the modern workplace has forced many employers to revisit their talent acquisition strategies, adopting new methods to secure the talent they need to succeed. While many focus on the external hunt for talent, there's an equally important strategy that often gets overlooked: empowering the talent you already have.

1. THE COST OF HIRING VS. THE VALUE OF EMPOWERMENT

Let's start with some facts. The hard costs of hiring include talent sourcing, selection, and onboarding, which average about $4,700 per employee, according to the latest benchmark from the Society for Human Resource Management. On top of that, it takes around 44 days to fill a position. These numbers highlight the significant investment involved in bringing new talent on board—an investment that doesn't even guarantee you've found the right fit.

So why do so many founders and CEOs hesitate to make the same investment in their existing team members? Why do they impose bureaucratic timelines for compensation, promotions, and empowerment, delaying the recognition and reward of talent that's already proven its worth?

If someone on your team deserves it, give it to them. Don't wait for some arbitrary timeline or simply because it feels too soon. If an employee has been with you for just a few months but has demonstrated outstanding character and work ethic, don't hold back. Empower them, reward them, and in return, you'll earn their loyalty, trust, and commitment to your vision.

2. EMPOWERMENT AS A CONTINUOUS PROCESS

Empowering your team doesn't have to be a daunting, time-consuming process. With the right structure in place, it can be less demanding and far more rewarding. Here are a few strategies to help you empower your team effectively:

- **Monthly Training Sessions:** Host a monthly training session for your team on a new topic relevant to their roles. This not only helps them develop new skills but also shows them that you're invested in their growth.
- **Quarterly Team Activities:** Organize a quarterly team activity and take everyone out for lunch or dinner. This fosters team bonding and creates a positive work culture where employees feel valued and appreciated.
- **Personalized Training Plans and KPIs:** Develop individualized training plans and Key Performance Indicators (KPIs) for each team member. This demonstrates that you recognize their unique contributions and are committed to helping them succeed.

Empowerment isn't just about giving out promotions or raises—it's about creating an environment where your team feels supported, valued, and motivated to do their best work. By making empowerment a continuous process, you build a culture of growth and development that benefits both your employees and your business.

3. THE RIPPLE EFFECT OF EMPOWERMENT

When you empower your team, you're not just investing in their success—you're investing in the success of your entire organization. Empowered employees are more engaged, more productive, and more likely to stay with your company for the long term. They become advocates for your brand, contributing to a positive work culture and helping to attract new talent.

Empowerment creates a ripple effect. When you give your team the tools, resources, and recognition they need to thrive, they'll empower you in return. They'll take ownership of their roles, drive your business forward, and help you achieve your goals.

LESSON 10: THE POWER OF WORDS

I get it, and I've been there. Startups are incredibly demanding, and sometimes, in the rush to keep everything moving, it's easy to overlook the small things that can make a big difference. When you're deeply invested in your business, it can start to feel like you're the one carrying the weight, like you're doing everyone else a favor—especially your employees. You might think that because of your relentless efforts, the company is growing, and everyone should be grateful to you.

While that might be true to some extent, it's crucial to remember that this is your business, not your employees' startup. They're helping you build your dream, and they deserve recognition for their contributions. This is where the power of simple compliments and kind words comes in—a tool that's often underestimated but can have a profound impact on your team.

1. THE IMPACT OF SIMPLE WORDS

Simple phrases like "Well done," "I'm proud of you," "We achieved this because of you," "Your efforts made this happen," and "Thank you for your hard work" might seem small, but they carry immense weight. These words are free, they take seconds to say or write, and they can significantly influence your company's culture and morale.

Imagine dropping a quick "Thank you" message on Slack or in a team chat after a project is completed. That simple act can uplift your team members, making them feel noticed and appreciated. It might even motivate them to push harder and take more ownership of their roles. As a leader, it's easy to focus on what needs fixing or what went wrong, but it's equally important—if not more so—to recognize what's going right and celebrate those wins.

2. WORDS AS A LEADERSHIP TOOL

As a leader, your words carry a lot of power. They can either uplift or discourage, inspire or demotivate. It's not just about finding and addressing concerns—it's about finding and acknowledging what's going well. Celebrating successes, no matter how small, can have a ripple effect throughout your team, boosting morale and productivity.

Don't just look for areas where you can provide feedback for improvement; also look for opportunities to express gratitude. Recognize the efforts of your team members publicly when appropriate. Acknowledge their contributions in meetings, in group chats, or even one-on-one. These small gestures can build a positive work environment where everyone feels valued.

3. THE POWER OF KIND WORDS: A STORY OF TWO PLANTS

To illustrate the power of kind words, I want to share a story that has always resonated with me. It's an experiment conducted by IKEA to demonstrate the impact of words on living beings—even plants.

In 2018, leading up to Anti-Bullying Day on May 4th, IKEA conducted an experiment called "Bully A Plant" at a school in the United Arab Emirates. The idea was simple: set up two identical plants in the school, and for 30 days, have students compliment one plant and bully the other. The plants were kept under identical controlled environments—they received the same amount of light, nutrition, and water. The only difference was the words directed at them.

The students recorded their words of praise for one plant and insults for the other, and these messages were played through speakers rigged into each plant's enclosure. After 30 days, the results were striking. The plant that received compliments was healthy and thriving, while the plant that had been subjected to negative comments was wilted and noticeably droopy.

While the experiment wasn't scientifically rigorous, it powerfully illustrates the impact that words can have. If words can do this to plants, imagine their effect on people.

4. EMPOWER YOUR TEAM WITH WORDS

The lesson here is simple: don't underestimate the power of kind words. Compliments and positive reinforcement are essential tools for any leader. They cost nothing, take little time, and can have a huge impact on your team's morale, productivity, and overall job satisfaction.

So, make it a priority to recognize and celebrate the good in your team. Don't wait for a special occasion—do it regularly, and watch how it transforms your work environment. Words have the power to uplift, inspire, and create a positive, thriving workplace. Use them wisely.

LESSON 11: WHY FOUNDERS SHOULD LISTEN

As a founder or CEO, it's easy to fall into the trap of thinking that your opinion matters most and that you know what's best for the business more than anyone else. After all, you're the one steering the ship, making the big decisions, and driving the vision forward. It can feel natural to give orders and delegate tasks around, assuming that you've got everything figured out. Yet, here's the truth: even when you think you know it all, taking a second opinion can be invaluable.

1. YOUR PERSPECTIVE ISN'T THE ONLY ONE

You don't run the entire company alone, and while you operate at the executive level, your line staff and employees are on the ground, experiencing things that you might not see on a daily basis. Let's say you need to make a decision related to client interactions, project management, or operational processes. In these cases, your employees might actually know more than you do because they are the ones directly involved in these areas. Taking their opinions into account could provide insights that you might overlook from your vantage point.

2. THE POWER OF DIVERSE PERSPECTIVES

You might be surprised by the suggestions and ideas that come from your team—ideas you hadn't considered before. While you have a certain perspective shaped by your experience and vision for the company, someone from your team might have a different angle that could be incredibly insightful. Hearing them out allows you to make more informed decisions, blending your strategic oversight with their practical experience.

3. THE MOTIVATIONAL BOOST OF BEING HEARD

Even if you ultimately decide not to act on someone's suggestion, simply asking for their opinion can have a powerful effect. It makes them feel noticed, valued, and respected. This feeling can significantly boost their productivity and motivation. When you show that you care about your team's input and that you're not too proud to listen, you foster a culture of openness and collaboration.

4. KNOWING WHEN TO SEEK OPINIONS

Of course, not every decision warrants a roundtable discussion. Some decisions need to be made quickly and don't require input from the entire team. However, when a decision involves a specific department, consider consulting with the team members who are directly affected. For decisions that involve multiple departments or the entire company, it might be worth having a roundtable discussion where everyone can pitch their ideas.

This is especially true for small startups with 10-20 employees, where you can maintain close relationships with your team. In such environments, a horizontal management structure—where communication flows more freely across all levels—can be faster, more effective, and more productive than a rigid vertical hierarchy.

5. KEEP IT HORIZONTAL

In a small startup, there's no need to strictly adhere to a top-down approach. Keeping things horizontal allows you to stay close to your team, understand their challenges, and leverage their insights. It also creates an environment where everyone feels they have a voice and that their contributions matter. This kind of inclusive leadership not only strengthens your decision-making process but also builds a stronger, more cohesive team.

In the end, being a great leader isn't just about having the right answers—it's about knowing when to ask for them. By valuing the perspectives of those around you, you're not only making better decisions but also creating a culture of respect, collaboration, and continuous improvement. So, even when you think you know it all, take a moment to seek a second opinion. You might be surprised by what you learn.

LESSON 12: NEGOTIATE TO WIN TOGETHER

One of the most valuable lessons I've learned in business negotiation comes from Bob Iger, the CEO and Chairman of The Walt Disney Company. Bob Iger has crafted some of the biggest deals in the history of business, including Disney's acquisitions of Pixar, Marvel, and the merger with 21st Century Fox. His approach to negotiation is not just about securing a win for Disney—it's about creating a situation where both parties walk away as winners. This is a crucial lesson for every entrepreneur, even if you're not running the largest entertainment company in the world.

1. THE TRUE NATURE OF NEGOTIATION

Young entrepreneurs might be influenced by movies like *The Wolf of Wall Street*, where sales and negotiation are portrayed as cutthroat games of domination, where the goal is to win at all costs, often at the expense of the other party. Yet in reality, effective negotiation is not about winning alone. It's about finding a middle ground where both parties can come out ahead, adding value to each other and setting the stage for a successful and sustainable partnership.

Negotiation should never be about ripping off a client or forcing someone into a deal that only benefits you. The ultimate goal is to close the deal in a way that leaves everyone satisfied, with both sides feeling like they've gained something valuable.

2. BE CANDID AND EXPEDIENT

One of the key lessons from Bob Iger's approach to negotiation is the importance of being candid. "I typically like to put things on the table in a fairly candid manner, in a direct manner, and quickly," Bob says. For him, expediency in dealmaking is paramount, and the best way to achieve it is through honesty and transparency.

To some, this might seem like a risky move—laying all your cards on the table upfront. Bob, however, believes that being upfront with your needs is crucial. Not only does it save time by cutting out the gamesmanship often associated with negotiation, but it also gives the other party a clear understanding of what you're looking for, while giving them the space to articulate their own needs.

This candid approach requires letting go of the zero-sum-game mentality. Instead of viewing the negotiation as a win-lose scenario, you open yourself up to the possibility that both parties can benefit significantly. This mindset shift can lead to more meaningful, long-term business relationships.

3. NEGOTIATING WITH A PARTNERSHIP MINDSET

Bob Iger's negotiation strategy reflects a partnership mindset. By being candid and focusing on mutual benefits, he ensures that the outcome is not just a transactional win, but the beginning of a potentially fruitful and ongoing relationship. This is a critical distinction—negotiation isn't just about closing a deal; it's about laying the foundation for future collaboration and success.

As an entrepreneur, adopting this mindset can transform the way you approach negotiations. It's not about trying to outmaneuver the other party or walking away with the biggest slice of the pie. It's about ensuring that both sides feel valued and that the agreement you reach is sustainable and beneficial in the long run.

4. THE REAL WIN IS IN THE CLOSURE

At the end of the day, the real victory in negotiation is closing the deal—not just winning the argument. By focusing on creating a win-win situation, you increase the likelihood of closing the deal in a way that is positive for everyone involved. This doesn't mean compromising your values or settling for less than what you deserve; it means finding a path forward that respects both parties' needs and objectives.

Remember, negotiation is not about defeating the other side—it's about building something together. Whether you're negotiating a contract, a partnership, or any other business deal, approach it with the mindset of collaboration and mutual benefit. Lead with candor, seek out common ground, and always aim to close the deal with both sides feeling like winners.

LESSON 13: ASSEMBLE YOUR A-TEAM

Building a successful business is not a solo endeavor. At some point, every entrepreneur needs to start assembling a team—a group of individuals who will help turn the vision into reality. Here's the catch: not just any team will do. To truly succeed, you need to assemble your A-Team.

1. QUALITY OVER QUANTITY

When it comes to hiring, it's tempting to think that more hands on deck will automatically lead to better results. In reality, the quality of your team members far outweighs the quantity. It's better to have a small team of top-tier talent than a large group of average performers. Every person you bring on board should elevate your business, not just fill a seat.

This means taking the time to carefully vet candidates, even if it slows down your hiring process. Rushing to fill positions can lead to costly mistakes. According to the U.S. Department of Labor, the average cost of a bad hire is 30% of the employee's first-year earnings. That's a price most startups can't afford to pay.

2. HIRE PEOPLE WHO ARE SMARTER THAN YOU

One of the smartest moves you can make as a leader is to hire people who are smarter than you. Yes, you read that right. Surround yourself with individuals who bring skills, knowledge, and perspectives that you don't have. These are the people who will challenge you, push your business forward, and bring ideas to the table that you might never have considered.

Hiring people who are smarter than you isn't about ego; it's about building a team that complements your strengths and fills in your weaknesses. If you're the smartest person in the room, you're in the wrong room. Your job as a leader is to assemble a team of experts who can take your business to the next level.

3. COMPLEMENTARY STRENGTHS MATTER

When building your A-Team, look for people whose strengths complement yours. If you're a visionary but struggle with execution, hire someone who excels at turning ideas into action. If you're a numbers person but lack creative flair, bring on someone with a strong creative background. The goal is to create a well-rounded team where everyone's strengths are leveraged, and weaknesses are covered.

This approach not only makes your team more effective but also fosters collaboration and innovation. When team members bring different skills and perspectives to the table, they can challenge each other, leading to better solutions and more innovative ideas.

4. INVEST IN YOUR PEOPLE

People are one of your greatest investments in business. When you hire the right individuals, you're not just filling a role—you're investing in the future of your company. That's why it's crucial to provide the support, resources, and opportunities they need to grow and thrive.

Invest in ongoing training and development, offer mentorship opportunities, and create a culture where people feel valued and empowered. Remember, a team that feels supported and appreciated will be more motivated, more productive, and more committed to your business's success.

5. THE COST OF HIRING THE WRONG PEOPLE

While hiring the right people can propel your business to success, hiring the wrong people can quickly take you down. A bad hire doesn't just affect the role they're in—it can have a ripple effect throughout your entire organization. It can lead to decreased morale, lower productivity, and even damage your company's reputation.

A study by CareerBuilder found that 74% of employers have made a bad hire, and of those, 37% reported that it negatively affected employee morale, and 18% said it hurt client relationships. The stakes are high, especially for startups where every team member plays a crucial role.

6. TAKE YOUR TIME, BUT DON'T HESITATE

Hiring the best people takes time, and that's okay. It's better to take the time to find the right fit than to rush and end up with someone who isn't aligned with your company's values or goals. However, once you've identified the right candidate, don't hesitate. Top talent doesn't stay on the market for long, so when you find someone who fits your needs, act quickly to bring them on board.

7. THE BOTTOM LINE: BUILD YOUR A-TEAM WISELY

In the world of business, your team is one of your most valuable assets. The right people can take your business to new heights, while the wrong ones can derail your progress. So, be deliberate in your hiring process. Look for individuals who are smarter than you, whose strengths complement yours, and who align with your company's vision and values.

Remember, assembling your A-Team isn't just about filling roles— it's about building a group of individuals who will drive your business forward, challenge you to be better, and help you achieve your long-term goals. Invest in your people, support their growth, and watch as they help propel your business to success.

LESSON 14: SUCCESS GROWS FROM EMPOWERED PEOPLE

A thriving business isn't just about the bottom line—it's about the people who make that success possible. Just like plants in a garden, your business and your employees need proper care and attention to grow. From my experience, the happier your employees are, the more successful your business will be. It's a simple equation that yields powerful results.

1. THE POWER OF HAPPY EMPLOYEES

Happy employees aren't just a nice-to-have—they're a cornerstone of a successful business. When your team feels valued, supported, and engaged, they're more productive, more creative, and more committed to your company's goals. They take pride in their work and go the extra mile because they believe in what they're doing.

However, this doesn't happen by accident. It requires deliberate effort and attention to create an environment where your employees can thrive. It's about more than just providing a paycheck—it's about building a culture where people feel empowered to do their best work.

2. THREE ESSENTIALS TO BOOSTING EMPLOYEE HAPPINESS

From my perspective, there are three key elements to creating a happy, empowered workforce:

- **Ensuring Financial Stability:** Financial stability is fundamental. If your employees are constantly worried about making ends meet, they won't be able to fully focus on their work. Offering fair and competitive wages is the first step in ensuring that your team feels secure and valued. It's not just about meeting the market rate—it's about recognizing the value that each person brings to your business and compensating them accordingly.
- **Hiring the Right People:** Building a great team starts with hiring the right people. It's not just about skills and experience—it's about finding individuals who align with your company's values and culture. When you hire people who are passionate about what they do and who fit well with your team, you create a positive, collaborative environment where everyone can succeed. Remember, one wrong hire can disrupt the entire team dynamic, so take the time to find the right fit.
- **Allowing for Work-Life Balance:** Work-life balance is more than just a buzzword—it's essential for maintaining employee happiness and preventing burnout. While the entrepreneurial journey often demands long hours, it's important to recognize that your team needs time to recharge and pursue their personal lives. Offering flexible working arrangements, respecting personal time, and encouraging employees to take breaks are all ways to support their well-being. When employees feel balanced, they're more energized and motivated at work.

3. TENDING TO YOUR BUSINESS AND YOUR PEOPLE

Think of your business and your employees' happiness as plants in a garden. Without proper care, they won't grow. Regularly check in with your team, listen to their needs, and be proactive in addressing any concerns. Just as a gardener waters and nurtures plants, you need to invest in the growth and well-being of your employees.

This also means being mindful of the work environment you create. Foster a culture of open communication, where employees feel comfortable sharing ideas, feedback, and concerns. Celebrate successes, learn from failures, and always show appreciation for the hard work your team puts in.

4. THE RIPPLE EFFECT OF EMPOWERMENT

When you empower your employees, the effects ripple throughout your entire business. Happy, engaged employees are more likely to provide excellent service to your customers, come up with innovative solutions, and contribute to a positive work culture. This, in turn, drives business success, leading to growth and profitability.

Moreover, when your employees feel valued and supported, they're more likely to stay with your company long-term. This reduces turnover, saving you the time and expense of constantly recruiting and training new staff. It also builds a sense of loyalty and commitment that's hard to match.

5. THE BOTTOM LINE: EMPOWERMENT EQUALS SUCCESS

Success in business isn't just about strategy, marketing, or finances —it's about people. By tending to your company and your employees with care, you create an environment where everyone can thrive. Remember, a happy team is a productive team, and when your employees are empowered, your business will grow.

Focus on ensuring financial stability, hiring the right people, and allowing for work-life balance. These elements are the foundation of a successful, sustainable business. When you take care of your people, they'll take care of your business—and that's a recipe for long-term success.

PART 3: MARKET NAVIGATION AND CUSTOMER FOCUS

LESSON 15: RIDE THE TREND, DON'T BECOME IT

In the fast-paced world of business, trends can seem like a goldmine —an opportunity to drive revenue, capture attention, and fuel growth. While trends are indeed important, here's a critical lesson: a business built solely on a trend is set to fail when that trend inevitably fades.

Trends come and go, often as quickly as they arrive. They can bring short-term success, but they rarely provide a stable foundation for long-term growth. I'm not saying you should ignore trends or fear change—doing so would mean missing out on valuable opportunities. However, there's a fine line between benefiting from a trend and becoming entirely consumed by it.

This was a lesson I had to learn from my mentor, Chaker Khazaal, who is also OBCIDO's co-founder and Executive Chairman. When we founded OBCIDO, I was eager to jump on the latest bandwagon, believing it was the fastest route to success. When NFTs exploded onto the scene, I wanted to pivot our entire marketing agency to focus purely on NFTs. I was convinced that this trend would replace traditional digital marketing, what we now refer to as Web 2.0.

Yet, Chaker had a different perspective. He advised against making such a drastic shift. Instead, we took a more balanced approach. We immersed ourselves in the NFT space, learning everything we could about this new trend—how it worked, how to market it, and how it fits within the broader digital landscape. We started offering Web 3.0 marketing services, but we didn't abandon our core digital marketing offerings.

The result? We saw a significant boost in revenue, at times achieving 5X growth during certain months, but unlike those who went all-in on the trend, our growth was sustainable. When the NFT craze began to wane, many companies that had fully committed to the trend struggled or failed. We, however, weren't affected. We had never lost sight of our broader business identity and goals.

As the market continued to evolve, we adapted. After NFTs, the focus shifted to content creation and short-form video content, and then to AI. Each time, we leveraged these trends to enhance our services and deliver more value to our clients, without losing our way. We remained true to our core mission as a startup, ensuring that trends complemented our business rather than defining it.

Here's the takeaway: trends are like waves—you should ride them, not become them. Benefit from the momentum they provide, but don't let them dictate your entire strategy. Keep your identity clear, your focus steady, and your overall business plan intact. By doing so, you'll be able to adapt to future trends with ease, ensuring that your business not only survives but thrives, no matter how the market shifts.

LESSON 16: YOU'RE NOT A DOCTOR

Building and growing a startup is no easy task. As an entrepreneur, I've often found myself overstressed and constantly worried, with an almost obsessive need to stay on top of everything. For a long time, I was addicted to checking my email. While most people check social media before they go to sleep and first thing when they wake up, I did the same—but with my inbox.

I remember one particular time when this behavior came into sharp focus. I was traveling to Paris and met up with a friend, Krisi. It was supposed to be a day off, a chance to enjoy the city, but as usual, I was out with my laptop and phone in hand. We sat down for lunch, and before I even looked at the menu, I opened my laptop to follow up on work.

That's when Krisi gently looked at me and said, "Omar, your work isn't a life-or-death situation. You're not a doctor, and your clients aren't patients waiting in an emergency room. They'll still be there if you take a moment to breathe."

Her words hit me hard. I was so focused on work that I hadn't realized how deeply I was letting it consume me. It was a wake-up call.

As entrepreneurs, we're conditioned to invest our time, energy, and sometimes our money into our startups. We pour ourselves into our work, often at the expense of everything else. Here's what I've learned: while it's crucial to be dedicated, it's just as important not to let that dedication turn into toxic stress.

1. DON'T INVEST IN STRESS

Stress, especially when it becomes chronic, doesn't serve you or your business. What's meant for you will come to you through hard work, honesty, and perseverance. What isn't meant for you won't be yours, no matter how much you stress over it. The key is to work hard and stay committed, but not at the cost of your well-being.

2. LEARNING TO LET GO

Ever since that day in Paris, I have noticed something. There were times when I worked incredibly hard for something, but it never materialized. Then there were other times when I decided to take a day off, to delay a project by a day or two, and you know what? Nothing catastrophic happened. Life continued as usual.

This realization was liberating. It taught me that, as an entrepreneur, it's okay to step back. It's okay to take a break. After all, I'm an entrepreneur, not a doctor. My clients won't face life-or-death situations if I don't respond immediately.

3. FINDING BALANCE

The entrepreneurial journey is demanding, but it doesn't have to be toxic. It's about finding balance between being fully invested in your work and knowing when to let go. It's about understanding that while your startup is important, so is your health, your peace of mind, and your ability to enjoy life.

This lesson is one I wish I had learned earlier, and it's one I hope every entrepreneur takes to heart. Work hard, stay honest, and do your best, but don't let your entrepreneurial drive turn into a source of constant stress. Your startup will thrive, not because you're always available, but because you've built it on a foundation of smart work, balance, and well-being.

Remember, you're an entrepreneur—not a doctor.

LESSON 17: BE YOU-CENTRIC, NOT ME-CENTRIC

One of the first lessons in communication, especially in the world of marketing, is the importance of being you-centric, not me-centric. This principle is foundational to value-based marketing, a customer-centric strategy that focuses on meeting your customers' needs rather than just promoting your own achievements.

1. UNDERSTANDING VALUE-BASED MARKETING

Value-based marketing is all about putting the customer at the center of your business strategy. Instead of marketing your company by highlighting your own greatness, you focus on your customers' wants and needs. It's about connecting with them on a deeper level and building trust by showing that you understand their pain points and have the solutions they're looking for.

For example, rather than boasting about your company's awards and certifications, shift the focus to your customer. Make them the hero of your story. Address their challenges, show empathy for their struggles, and provide clear, data-backed evidence of how your product or service can add value to their lives.

Simply put, a value-based marketing campaign helps you identify what your customers are struggling with and positions your product as the solution to their problems. This approach not only resonates more with your audience but also builds a stronger, more trusting relationship between your brand and your customers.

2. THE IMPORTANCE OF CUSTOMER-CENTRIC COMMUNICATION

Even if you're not running a marketing agency, the need for effective marketing, branding, and sales is universal in any business. Communication is key, no matter the industry, and the most effective communication is you-centered.

When communicating with your audience—whether in marketing materials, sales pitches, or customer service—you must never be me-centered. Instead, focus on being you-centered. Ask yourself: how can I make this about the customer? How can I speak to their needs, address their concerns, and show them that I'm here to help them succeed?

A simple yet common mistake many brands make is showcasing their achievements right at the top of their homepage. While these accolades are impressive, they don't directly add value to your customer. What does add value is addressing their specific needs and demonstrating how your business can solve their problems. Once you've established that connection, you can subtly weave in your credentials to reinforce your credibility.

3. APPLYING THE YOU-CENTRIC APPROACH ACROSS CHANNELS

The you-centric approach isn't just for marketing; it should be applied across all areas of your business. Whether you're writing an email, creating a presentation, or engaging on social media, always think about your audience first. Make them the focal point of your communication.

When it comes to your website, for example, the hero section should immediately speak to your customer's challenges and how you can help them overcome these obstacles. Use clear, relatable language that resonates with their needs and aspirations, rather than leading with your own accomplishments. This approach not only captures attention but also builds an instant connection that encourages further engagement.

4. BUILDING LASTING CONNECTIONS

By adopting a you-centric approach, you're not just selling a product—you're building a relationship. Customers want to feel understood and valued. When you center your communication around them, you create a deeper connection that goes beyond a simple transaction. This approach fosters loyalty, encourages repeat business, and can turn customers into advocates for your brand.

In the end, being you-centric is about shifting your mindset from "How can I promote my business?" to "How can I help my customers?" This subtle but powerful change in perspective can transform your marketing efforts and lead to stronger, more meaningful connections with your audience.

LESSON 18: SOCIAL MEDIA IS NOT THE MEASURE OF SUCCESS

In today's world, it's easy to get caught up in the social media frenzy, especially as an entrepreneur. We're constantly bombarded with images and stories of so-called "successful" individuals flaunting their wealth, lifestyle, and hustle. However, here's the truth: social media is often more fiction than fact. As an entrepreneur, your worth should never be judged by your personal brand or online presence.

Let me be clear—I'm not accusing every content creator of being illegitimate. I also create content to promote our brands, but I approach it differently. My goal is to add value with my content, to educate, share marketing tips, tools, strategies, and real marketing stories. What I don't do is sit behind a camera and school people on how hard they should work while belittling every 9-to-5 employee.

Because here's another truth: whether you choose to be an entrepreneur or work a 9-to-5 job, both paths are valid, and both can be fulfilling. It all depends on your character, your goals, and your preferences, but that's a topic for another lesson. For now, let's get back to the real lesson here.

1. THE SILENT TITANS OF ENTREPRENEURSHIP

Have you heard of Melanie Perkins? Chances are, you haven't. Yet she's the co-founder and CEO of Canva, with a net worth of $4.4 billion. What about Eric Yuan? He's the founder of Zoom, with a net worth of $3.8 billion as of August 2024. Ben Chestnut? He's the co-founder of Mailchimp, with a net worth of $4.9 billion.

Here's the kicker: Melanie Perkins has just around 35,000 followers on Instagram. Eric Yuan doesn't even have an Instagram account, and Ben Chestnut has fewer than 3,000 followers. Meanwhile, their companies—Zoom, Canva, and Mailchimp—are household names in the tech and marketing industries. I bet there's no one in these fields who hasn't heard of them.

On the other hand, think about the social media influencers and content creators you follow. Can you name any companies they've built from the ground up? Probably not. Why is that? Because social media doesn't always reflect reality.

2. DON'T GET DRAGGED INTO THE SOCIAL MEDIA TRAP

It's easy to feel like you're falling behind when you see other entrepreneurs going viral on social media. However, here's a critical lesson: just because someone has a strong social media presence doesn't mean they're ahead of you in the real world. Some of the world's best entrepreneurs have built billion-dollar businesses with minimal or even negligible social media presence.

Social media can be toxic if you let it dictate your self-worth or your business goals. It's not a measure of success, and it's certainly not a metric that defines your value as an entrepreneur. Don't compare your journey to someone else's highlight reel. Instead, focus on what truly matters—building a strong, sustainable business that solves real problems and delivers real value.

3. CHOOSE YOUR CONTENT WISELY

The next time you find yourself getting sucked into the social media vortex, remind yourself that not everything you see online is true. Don't get dragged down by comparisons or the pressure to keep up with the latest trends. Instead, be selective about the content you consume.

Follow creators who add value to your life, who teach you something useful, and who provide insights that help you grow in your entrepreneurial journey. Seek out content that educates you, challenges you, and pushes you to be better—not content that makes you question your worth or your progress.

Remember, the most successful entrepreneurs aren't necessarily the ones with the most followers or the flashiest posts. They're the ones who quietly build billion-dollar businesses while staying true to their vision, their values, and their purpose. Keep that in mind the next time you're tempted to judge yourself by your social media presence—or anyone else's.

LESSON 19: STUDY THE MARKET, DON'T MEMORIZE THE COMPETITION

In the startup world, it's easy to get caught up in what everyone else is doing, especially when you're running a business in a crowded space like marketing. Unless you're Coca-Cola competing with Pepsi or Apple going head-to-head with Samsung, your focus shouldn't be on what your competitors are doing—it should be on what you're doing.

When we first co-founded OBCIDO, it was tempting to compare ourselves to the countless established agencies that had been around for years. As of January 29, 2024, there were over 433,400 advertising agencies worldwide. With so many players in the game, it's easy to feel like your individual chances of success are minuscule —something like 0.000002%. If you spend your time obsessing over every competitor out there, you'll end up overwhelmed, distracted, and ultimately, stagnant.

1. THE TRAP OF OVER-FOCUSING ON COMPETITORS

Many startups make the mistake of looking at established businesses and trying to replicate their success. They see what other companies are doing in sales, marketing, and product development, and they try to follow suit, thinking that it's the only way to get ahead. This approach, however, can backfire. Not only can it lead to a lack of originality, but it can also cause you to invest in strategies that might not be right for your business.

At OBCIDO, we quickly realized that focusing too much on the competition was counterproductive. Yes, it's important to understand the market and be aware of industry trends, but memorizing every move your competitors make only leads to second-guessing and hesitation. Instead, we chose to focus on our strengths and carve out our own path.

2. FINDING OUR OWN WAY

Rather than trying to do everything or mimic every successful agency, we honed in on our core services. We didn't chase after every new trend or try to be everything to everyone. Instead, we specialized in a few key areas, listened to our clients, and adapted our offerings based on their needs. This approach allowed us to develop our unique market share and grow steadily.

We also avoided the trap of comparing our progress to that of other agencies. While some might be making millions more, we're proud of the fact that we achieved over $1M in yearly revenue by our third year. We did this by staying true to our vision and not getting distracted by what others were doing. In a world where only 25% of new businesses make it to 15 years or more, and 90% of startups fail, building a fully bootstrapped business that's thriving is something we're incredibly proud of.

3. THE POWER OF FOCUS

Success doesn't always mean being the biggest or the fastest-growing agency out there. It's about building something sustainable, something that reflects your unique strengths and vision. By focusing on what we do best and not getting lost in the competition, we've been able to create a business that's not only successful but also fulfilling.

In an industry where new agencies are popping up every day, it's easy to get caught up in the noise. Just remember, each business's journey is unique. The key is to stay focused on your own path, keep refining your offerings, and build something that's true to you and your clients.

In the end, the lesson is simple: study the market, but don't memorize the competition. Keep your focus on what you do best, stay true to your vision, and build a business that reflects your unique strengths. Success isn't about being the best in the industry —it's about being the best version of your own business.

LESSON 20: LEAD WITH THE 'WHY' BEFORE THE 'WHAT'

When it comes to leadership and communication, one of the most powerful tools you can use is leading with the "why" before the "what." Whether you're implementing a change, addressing a mistake, or expressing gratitude, starting with the why helps people understand the bigger picture and connect emotionally with the message.

1. THE POWER OF 'WHY'

People tend to relate to and remember the why more than the what. The why is often associated with results, reasons, and purpose, whereas the what merely states the facts. Great leaders and successful companies inspire action by starting with why—not just what they want you to do or what product they want you to buy, but why it matters. They tap into emotions, understanding that emotion is a far greater motivator of action than rational thought.

When you lead with the why, you're connecting with your audience on a deeper level. You're not just telling them what happened or what needs to be done—you're explaining the significance behind it. This approach is not only more persuasive but also more memorable.

2. PUTTING 'WHY' INTO ACTION

Let's look at how this works in practice:

- **Addressing a Mistake:** Suppose an employee was late in sending an important email to a client. Instead of focusing solely on what happened (the delay) and what the disciplinary action might be, start with the why. Explain why timely communication is crucial for client satisfaction and the continuity of the business. A delayed email can disrupt the client's workflow, leading to missed deadlines, client frustration, potential contract termination, and ultimately, a negative impact on the company's revenue and reputation. By understanding the why, the employee can better grasp the importance of their actions and how it affects the bigger picture.
- **Recognizing Exceptional Work:** On the flip side, when an employee delivers exceptional work, closes a deal, or goes the extra mile, don't just acknowledge what they did —highlight why it matters. Explain how their efforts contribute to the success of the company, strengthen client relationships, or open new opportunities for growth. When employees understand the impact of their contributions, they feel more valued and motivated to continue performing at a high level.
- **Implementing Change:** When you're making changes within the company, always start with why the change is necessary before diving into what will change. For instance, if you're restructuring a team, explain why the new structure will improve efficiency, enhance collaboration, or better align with the company's goals.

When people understand the reasons behind the change, they're more likely to support it and adapt smoothly.

3. WHY THE 'WHY' MATTERS

Leading with the why isn't just about effective communication— it's about fostering a culture where everyone understands the purpose behind their work. When your team knows why their actions matter, they're more engaged, more committed, and more aligned with the company's mission.

The why provides context, creates meaning, and drives motivation. It turns routine tasks into purposeful actions and helps people see how their individual contributions fit into the larger narrative of the company's success. Whether you're leading a team, launching a product, or simply giving feedback, always start with the why.

In the end, leading with the why is about more than just getting things done—it's about inspiring action, building trust, and creating a shared sense of purpose. So, the next time you're communicating with your team, remember to lead with the why before the what. It's a small shift that can make a big difference.

LESSON 21: MAKE VALUE YOUR MISSION

Let's talk about a common goal that nearly everyone has: to be successful. It sounds great, right? Yet here's the thing—chasing success is like chasing the wind. The harder you chase, the more elusive it becomes. Instead, what if you shifted your focus? What if, instead of making success your goal, you made being valuable your mission? When you become valuable, success doesn't need to be chased—it's naturally attracted to you.

1. THE POWER OF BEING VALUABLE

Success is often seen as the ultimate destination, but value is the journey that gets you there. Think about it: the most successful people, businesses, and brands aren't those who constantly strive for success. They're the ones who consistently provide value. They solve problems, fulfill needs, and bring something to the table that others can't.

When you make value your mission, you're not just working toward success—you're creating something that others genuinely need and appreciate. This approach shifts the dynamic. Instead of you chasing opportunities, opportunities start coming to you because people recognize the value you bring.

2. BE VALUABLE, NOT JUST AVAILABLE

There's a crucial distinction between being valuable and being available. While it might seem like being available all the time would increase your value, the opposite is often true. The less available you are, the more valuable you become.

This is especially true for consultants, freelancers, and anyone in a service-based industry. If you're always available, people may start to take you for granted. Your time and expertise become less of a sought-after resource and more of an expectation. However, when you position yourself as a valuable expert whose time is limited, you increase your perceived value. Clients and customers begin to see you as someone worth waiting for, someone whose insights and skills are in high demand.

3. QUALITY OVER QUANTITY

Being valuable isn't about spreading yourself thin across as many projects, clients, or tasks as possible. It's about focusing on the quality of what you offer. It's better to deliver exceptional value to a few clients than to provide mediocre service to many. When you focus on quality, you build a reputation for excellence, and that's something people are willing to pay a premium for.

Take a cue from the world's top luxury brands. They don't aim to be available to everyone; they aim to be valuable to the right people. They understand that exclusivity and scarcity increase value. In the same way, by focusing on delivering top-notch value rather than being available to everyone at all times, you position yourself as a premium option in your field.

4. MAKE VALUE YOUR BRAND

Value should be at the core of your brand identity. It's not just about what you do; it's about how you do it and the impact you have on others. Whether you're building a business, developing a product, or offering a service, your goal should always be to create value that resonates with your audience.

Ask yourself: What problem am I solving? How am I making life easier, better, or more enjoyable for my customers? What unique value do I bring to the table that no one else does? When you can answer these questions clearly, you're on the right path to making value your mission.

5. VALUE ATTRACTS SUCCESS

When you focus on being valuable, success becomes a byproduct of your efforts. It's not something you have to chase because it naturally follows value. People are drawn to those who provide solutions, offer insights, and genuinely make a difference. By making value your mission, you create a magnetic pull that attracts success, opportunities, and loyal customers.

In a world where everyone is vying for attention, being valuable sets you apart. It makes you stand out in a crowded marketplace and positions you as a leader in your field. Whether you're an entrepreneur, consultant, or employee, making value your mission will elevate your work and bring you the success you seek.

6. THE BOTTOM LINE: VALUE FIRST, SUCCESS SECOND

So, here's the takeaway: don't make success your primary goal. Make value your mission. When you focus on delivering value—whether through your products, services, or personal brand—success will naturally follow. Remember, being valuable doesn't mean being available 24/7. It means offering something so good, so impactful, that people are willing to wait for it, seek it out, and pay for it.

In the end, the most successful individuals and businesses are those who understand that value is the true currency of success. So, make value your mission, and watch as success comes knocking on your door.

PART 4: EXECUTION AND OPERATIONAL EFFICIENCY

LESSON 22: GENEROSITY WITHOUT LOSING YOUR VALUE

As the CEO of a four-year-old startup with over 100 clients globally, I've learned that there's a fine line between being generous and devaluing yourself. This is one of those lessons that's easy to understand in theory but challenging to grasp in practice—especially when you're navigating the pressures of building a business from the ground up.

Generosity can be a powerful tool. It builds goodwill, strengthens client relationships, and can set you apart from competitors. Nevertheless, there's a catch: generosity needs to be balanced with a clear sense of your own value. Without that balance, you risk not only undermining your business but also eroding your self-worth as an entrepreneur.

Here's how I've learned to navigate this delicate balance:

1. TEST THE WATERS

When you're starting out, it's natural to want to impress clients by going above and beyond. Being generous—whether it's through extra services, discounts, or added value—can leave a lasting positive impression, but it's crucial to test the waters first.

Give the benefit of the doubt and be generous the first time around. If the client appreciates your effort and values the extra mile you've gone, that's a win. However, if you notice that your generosity is taken for granted or, worse, expected as the norm, it's time to take note. Adjust your approach and set boundaries.

2. VALUE YOURSELF

The early stages of a startup can be particularly tricky when it comes to valuing your work. There's a constant push-pull between wanting to attract and retain clients and ensuring that you're not underselling yourself. But here's the truth: if you don't value your skills and your business, no one else will.

There's a thin line between being a generous entrepreneur who truly values their work and becoming a doormat for clients who don't respect what you bring to the table. Don't let your desire to please lead you to undervalue your services. Know your worth, set your rates accordingly, and be prepared to walk away from clients who don't respect that.

3. LEARN AS YOU GROW

When you're just starting out, it's easy to make mistakes—whether it's undercharging, over-delivering without recognition, or being overly generous with clients who don't reciprocate. However, every mistake is an opportunity to learn.

In the early days, I made plenty of these mistakes. I allowed clients to take advantage of my generosity, thinking it would eventually lead to better relationships or more business. However, I learned quickly that this approach is unsustainable. The key is to learn from these experiences, refine your approach, and minimize these situations in the future.

4. BALANCE GENEROSITY WITH BOUNDARIES

Generosity and value are not mutually exclusive. In fact, the most successful entrepreneurs understand how to balance the two. Be generous with clients who appreciate and respect your efforts, but don't be afraid to set firm boundaries with those who don't. Your time, expertise, and resources are valuable—treat them as such.

At the end of the day, it's all about balance. Be generous, but never at the expense of your own value. Over time, you'll develop a keen sense of when to give and when to hold back, ensuring that both your clients and your business are respected.

After four years as a CEO, I can tell you this: it's possible to be generous without losing your value. It's a delicate balance, but it's one that's essential for long-term success.

LESSON 23: THE FIRST STEP

Throughout my journey as an entrepreneur, I've faced countless projects and tasks that seemed overwhelming at first glance. There were hundreds of times when I looked at a project and thought to myself, "Where do I even begin?" or "How can I possibly bring this to fruition?" Here's what I've learned: the key to overcoming that initial overwhelm is simply to take the first step. Once you do, everything else tends to follow naturally.

There's something almost magical about taking that first step. No matter how daunting a project might seem, once you start, the pieces begin to fall into place. The hardest part is often just getting started. As soon as you take that initial action, the next step feels easier, and the one after that is easier still.

1. OVERCOMING THE OVERWHELM

I've felt overwhelmed more times than I can count, but each time, the process has been the same: start with the first step. It doesn't matter what you're working on—whether it's a massive project or a small task—the moment you take that first action, you've already begun to overcome the mountain in front of you.

Take, for example, a project I once managed: building an 80-page website for a major client. At first, the scope of the project felt enormous—how was I going to pull this off? Instead of letting the scale of the project paralyze me, I broke it down into smaller, more manageable tasks.

2. BREAKING DOWN THE BIG PICTURE

When you're faced with something that seems insurmountable, break it down. For that 80-page website, I broke the project into stages: strategy, content writing, design, design approval, development, content upload, testing, launching, and optimizing. Suddenly, the project wasn't one massive, intimidating task—it was a series of smaller tasks that I was already familiar with, tasks that I knew how to handle.

By focusing on completing one step at a time, I was able to move forward steadily, and before I knew it, the entire project was complete. Looking back, I'd often reflect, "Was this really the same project that once felt impossible? How did I navigate through such complexity?" The answer was always the same: I started with the first step.

3. KEEP MOVING FORWARD

The lesson here is simple but powerful: don't stress, even when things aren't clear. Just keep moving forward. The first step is always the hardest, but once you've taken it, you'll find that the path ahead becomes clearer and more manageable. No matter what you're facing, whether it's a large project, a new venture, or an unfamiliar challenge, just take that first step.

You'll always have worries and doubts—that's natural. However, if you can muster the courage to start, you'll find that the journey becomes easier with each step you take. The impossible starts to feel achievable, and before long, you'll be looking back, amazed at how far you've come.

Remember, the hardest part is getting started. Once you do, everything else will follow.

LESSON 24: YOU CAN'T CONTROL WHAT YOU CAN'T MEASURE

KPIs—Key Performance Indicators—are often seen as restrictive or cumbersome, especially in the fast-paced world of startups. Some might argue that KPIs force you to operate within the confines of specific metrics, making it difficult to adapt to the complexities and nuances of running a business. Yes, things can sometimes be complex and not easily measured, particularly when you're just starting out. Here's the reality: you can't control what you can't measure.

In today's business landscape, analytics has become a hot topic. Business analytics creates a solid foundation for informed decision-making, helping companies identify emerging trends, market opportunities, and potential risks. This results in more effective strategies that drive success and reduce uncertainties.

1. THE IMPORTANCE OF KPIS, EVEN FOR STARTUPS

While I won't dive deep into the science of business analytics in this lesson, it's crucial to understand that most growing companies invest heavily in analytics to measure their KPIs accurately and make data-driven decisions. But what does that mean for smaller companies and startups who may not have the resources to hire a data analyst?

KPIs stand for Key Performance Indicators—a quantifiable measure of performance over time for a specific objective. Think of KPIs as measurable goals that help you track progress and make adjustments as needed. They aren't just for big corporations; they're essential for startups too.

2. BREAKING DOWN KPIS: A PRACTICAL APPROACH

Let's make this concept easier to apply. Imagine KPIs as quantifiable targets for anything you do. They're not just for sales. While many founders focus on setting sales targets, they often overlook other crucial areas like spending, team performance, and productivity. Without KPIs in these areas, you risk losing control over your operations.

For example, if you don't set KPIs for spending, you might find yourself hemorrhaging money on unnecessary expenses. I've personally seen companies with thousands of dollars in SaaS (Software as a Service) subscriptions that are hardly used. Similarly, without KPIs for each department or employee, you won't be able to measure performance effectively, making it challenging to identify areas that need improvement.

3. WHAT SHOULD YOU MEASURE?

In the dynamic environment of a startup, knowing what to measure is key to maintaining control and driving progress. Here's a streamlined approach to setting KPIs across different areas of your business:

- **Sales KPIs:** Focus on measuring essential metrics like leads generated, market coverage, sales revenue, average customer lifetime value, and referrals. These indicators help you track the health and growth potential of your sales efforts.
- **Spending KPIs:** Keep a close eye on your expenditures by tracking salaries, subscriptions, and miscellaneous costs. This helps you avoid unnecessary spending and ensures your budget is being used effectively.
- **Team Performance KPIs:** Evaluate your team's output by assessing production levels, the quality of customer service, and employee growth. These KPIs offer insights into how well your team is performing and where improvements can be made.
- **Marketing KPIs:** Analyze the impact of your marketing campaigns by measuring growth metrics, reach, lead generation, and the increase in followers. This allows you to gauge the effectiveness of your marketing strategies and make necessary adjustments.
- **Productivity KPIs:** Measure productivity at various levels, including your team's overall efficiency, your personal productivity, the effectiveness of your processes, communication flows, and operational productivity. These indicators help you optimize workflows and improve overall efficiency.
- **Customer Success KPIs:** Track customer feedback, online reviews, word-of-mouth referrals, and your brand's reputation. These KPIs are vital for understanding customer satisfaction and ensuring long-term loyalty.

While it's true that startups often need to make quick, instinctive decisions, whenever possible, apply structured measurement. KPIs provide the control and insight necessary for making informed choices, aligning every aspect of your business with your broader objectives.

4. A PRO TIP FOR MANAGING SAAS SUBSCRIPTIONS

Now, let's talk about something I wish I'd known when I started my business. As you grow your startup, you'll likely subscribe to various services—website hosting, email subscriptions, Zoom, Notion, and countless other tools. Did you know that an average organization uses around 110 SaaS applications for its operations? And that 78% of small businesses have invested in SaaS solutions?

SaaS is a fantastic business model, but it's easy to fall into the trap of subscribing to numerous tools, using only a fraction of their capabilities, or worse, forgetting about them altogether. Before you know it, your credit card is being charged for 100 different SaaS tools, many of which you're barely using.

Here's a simple strategy to prevent this from happening: create an Excel or Google Sheet dedicated to tracking your SaaS subscriptions. Include columns for the SaaS name, website access link, username, password, payment method (last 4 credit card digits), subscription type (monthly, yearly, lifetime), and billing cycle. You can also invest in a tool like Rocket Money, which automates this process for you.

Good housekeeping from the start is crucial. By keeping track of your SaaS subscriptions and other expenses, you'll avoid wasting thousands of dollars on unutilized resources and maintain better control over your finances.

5. THE BOTTOM LINE

In business, especially in a startup, you can't control what you can't measure. KPIs provide a framework for accountability, improvement, and growth. Whether it's tracking sales, spending, productivity, or customer success, KPIs help you stay on course and make informed decisions that drive your business forward. So, don't shy away from setting KPIs—embrace them as a vital tool in your entrepreneurial toolkit.

LESSON 25: YOU REQUIRE LESS TO SUCCEED

In the fast-paced, competitive world of startups, it's easy to get swept up in the belief that you need to have everything perfectly in place from day one—a big team, a fancy office, the latest tools, and a hefty budget for client acquisition. Yet the truth is, you can build a successful business with far fewer resources than you might think. In fact, sometimes, less is exactly what you need to succeed.

1. LEAN TEAMS, GREATER AGILITY

There's a common misconception that a large team equates to more clients, faster results, and greater success. However, in the early stages of your business, a smaller, leaner team can often be more effective. With fewer people, communication flows more smoothly, decisions are made faster, and the culture feels more cohesive. Each team member ends up wearing multiple hats, which not only keeps overhead low but also fosters a sense of ownership and versatility that larger teams sometimes lack.

A lean team also allows for greater agility. In the constantly changing landscape of business, being able to pivot quickly is invaluable. When your team is small, you can adapt to new trends, technologies, and customer needs without the inertia that often slows down larger organizations. This flexibility can be a significant competitive advantage, especially in the early stages of your business's growth.

2. CUSTOMER RELATIONSHIPS OVER BIG BUDGETS

When it comes to growing your customer base, the strength of your relationships often outweighs the size of your marketing budget. While it's tempting to think that you need to spend big on advertising to attract customers, word-of-mouth, referrals, and networking can be far more powerful. Building strong, genuine relationships with your customers—understanding their needs, delivering exceptional results, and exceeding their expectations—will do more for your business's growth than any expensive ad campaign ever could.

Customers value trust and reliability. They want to know that you have their best interests at heart and that you'll go the extra mile to help them succeed. By focusing on delivering real value and fostering strong relationships, you'll not only retain customers but also benefit from the most effective form of marketing: word-of-mouth referrals.

3. START WITH WHAT YOU HAVE

In today's tech-driven world, it's easy to fall into the trap of thinking you need the latest tools and software to run a successful business. Still, the reality is that many successful companies started with just the basics. Rather than splurging on every new tool or platform, focus on being strategic with what you have. Prioritize your resources and invest in tools that are truly essential to your operations.

As your business grows, you can gradually invest in more advanced tools and technology. In the beginning, it's more important to have a clear strategy and a deep understanding of your customers' needs than to have every possible tool at your disposal. Remember, tools are just that—tools. They're only as effective as the strategy and execution behind them.

4. RETHINK OFFICE SPACE

The idea of having a prestigious office in a prime location is undoubtedly appealing, but in today's world, it's not a necessity. Many successful businesses began in shared coworking spaces, or even from home offices, where overhead costs are minimal. What customers care about isn't the view from your window or the size of your conference room—it's the quality of your work and the value you provide.

By keeping your overhead low, you can invest more in your team, in customer deliverables, and in the areas that truly drive your business forward. In a world where remote work is becoming increasingly common, the location of your office matters far less than the value you provide to your customers.

5. FLEXIBLE SERVICE OR PRODUCT OFFERINGS

You don't need to offer every product or service under the sun from day one. In fact, trying to do too much too soon can dilute your focus and stretch your resources thin. Instead, start with your core strengths—whether that's a specific product line, a niche service, or a unique offering—and gradually expand as you gain more experience and as customer demand grows.

This approach not only makes your operations more manageable but also positions you as a specialist in your key areas. Customers are often more attracted to businesses that excel in specific niches than to those that offer a little bit of everything but don't stand out in any particular area.

6. DIY WHERE YOU CAN

In the early stages of your business, taking on tasks like bookkeeping, IT management, and even basic design work yourself can save significant amounts of money. While these tasks might not be glamorous, they give you a better understanding of your business's inner workings and help you keep control over your expenses.

As your business grows, you can bring in specialists or outsource these tasks, but initially, DIY solutions are often more than sufficient. This hands-on approach not only saves money but also empowers you with a deeper knowledge of your business's operational needs.

7. ADAPTABILITY OVER RIGID STRUCTURES

In the dynamic world of business, change is the only constant. Whether it's market shifts, new technologies, or evolving customer needs, the ability to adapt quickly is crucial. A flexible, adaptable approach will serve you far better than rigid structures and processes.

By staying nimble and responsive, you can adjust your strategies to meet real-time challenges and seize new opportunities as they arise. This adaptability not only helps you stay ahead of the curve but also ensures that your business remains relevant and effective in a rapidly changing landscape.

8. BUILD SLOWLY, BUT STEADILY

Many successful businesses started small, with just a few key customers and a narrow focus. Instead of trying to grow too quickly, concentrate on delivering exceptional value to each customer and gradually building your reputation. A strong foundation built on consistent, high-quality work will lead to sustainable growth, rather than the boom-and-bust cycle that often comes from scaling too fast.

By focusing on steady, manageable growth, you can avoid the pitfalls of overexpansion and ensure that your business is built to last.

In the business world, success doesn't require a massive team, a huge budget, or a high-end office. By starting lean, focusing on your strengths, and prioritizing customer relationships and results, you can build a thriving business with far fewer resources than you might think. The key is to remain adaptable, stay focused on what truly matters, and let go of the notion that bigger is always better.

Sometimes, less is exactly what you need to succeed.

LESSON 26: SHORTER MEETINGS, BETTER RESULTS

In the business world, meetings have become a staple of daily operations. That said, let's be honest—how many times have you sat through a meeting that felt like it could have been an email? Or found yourself zoning out halfway through a long-winded discussion that seemed to go in circles? The reality is that meetings when not properly managed, can be one of the biggest time sinks in any organization. That's why I've learned to embrace a simple principle: shorter meetings lead to better results.

1. THE PROBLEM WITH LONG MEETINGS

Long meetings often lead to diminishing returns. The longer they drag on, the more likely participants are to lose focus, become disengaged, or get bogged down in unnecessary details. Instead of driving productivity, these meetings can stifle it, leaving everyone feeling drained rather than energized.

In my experience, a lengthy meeting often indicates a lack of clarity or purpose. When a meeting doesn't have a clear agenda or when participants are unclear about the desired outcomes, discussions tend to meander. What could have been a quick decision-making session turns into an extended debate, eating up valuable time that could be better spent on actual work.

2. THE POWER OF A CLEAR AGENDA

The first step to shortening meetings is to start with a clear, concise agenda. Before scheduling a meeting, ask yourself: What do I want to accomplish? What decisions need to be made? What input is required from the participants? By setting a clear agenda, you establish a roadmap for the meeting, ensuring that everyone is on the same page from the start.

A well-defined agenda not only keeps the meeting on track but also helps participants prepare in advance, allowing for more focused and productive discussions. It's also a good idea to share the agenda ahead of time, so everyone knows what to expect and can come prepared with their thoughts and ideas.

3. STICK TO THE ESSENTIALS

When it comes to meetings, less is more. Focus on the essential topics that require real-time discussion and decision-making. If an issue can be resolved via email or a quick chat, there's no need to include it in the meeting. Prioritize the most important topics and save the rest for another time—or better yet, handle them outside of the meeting altogether.

By limiting the scope of your meetings to what truly matters, you not only shorten their duration but also increase their impact. Participants are more likely to stay engaged and contribute significantly when they know their time is being respected and used wisely.

4. SET A TIME LIMIT—AND STICK TO IT

One of the most effective ways to ensure shorter meetings is to set a strict time limit. Whether it's 15 minutes, 30 minutes, or an hour, decide in advance how long the meeting will last and communicate this to all participants. Then, stick to it.

A time limit creates a sense of urgency that encourages participants to stay focused and on-topic. It also forces everyone to prioritize their input, ensuring that the most important points are discussed first. If time runs out and there are still outstanding issues, consider scheduling a follow-up meeting or handling them through other channels.

5. ENCOURAGE BREVITY AND FOCUS

During the meeting, encourage participants to be brief and to the point. Long-winded explanations and irrelevant discussions can quickly derail a meeting and extend its duration unnecessarily. Instead, ask everyone to focus on the key points, make their case concise, and avoid going off on tangents.

As a leader, it's your responsibility to steer the conversation back on track if it starts to veer off course. Don't be afraid to cut off discussions that are no longer productive or that delve too deeply into details that can be addressed later.

6. FOLLOW UP WITH ACTION ITEMS

A short meeting doesn't mean a lack of action. In fact, shorter meetings are often more effective because they leave participants with a clear sense of what needs to happen next. At the end of the meeting, summarize the key decisions made and outline the action items for each participant.

Follow up with a brief email reiterating these action items, along with deadlines and any necessary resources. This ensures that everyone knows their responsibilities and can move forward with clarity and purpose.

7. THE BENEFITS OF SHORTER MEETINGS

By embracing shorter meetings, you're not only reclaiming valuable time for yourself and your team but also fostering a more efficient and productive work environment. Shorter meetings lead to quicker decision-making, more focused discussions, and better overall results.

In today's fast-paced business world, time is one of your most valuable resources. By cutting down on lengthy, unproductive meetings, you can free up more time to focus on what really matters—growing your business, serving your customers, and achieving your goals.

Remember, the goal of a meeting is to drive action, not to fill time. The next time you schedule a meeting, challenge yourself to keep it short, focused, and impactful. You'll be amazed by the difference it can make.

LESSON 27: DON'T BE THE JACK OF ALL TRADES

When you're starting a business, it's tempting to try and do everything yourself. After all, it's your vision, your passion, and your responsibility. Here's the reality: you can't be an expert in everything, and trying to be will only hold you back. The key to building a successful business isn't mastering every skill—it's knowing when to bring in the professionals.

1. RECOGNIZE YOUR SKILLS GAPS

Every founder has their strengths, but they also have their weaknesses. Maybe you're great at sales and strategy, but you struggle with the financial side of things. Perhaps you excel at product development but find yourself lost when it comes to marketing. It's important to recognize where your skills gaps are because these are the areas where you're most likely to need help.

Trying to cover these gaps on your own can lead to mistakes, inefficiencies, and a lot of unnecessary stress. For example, if accounting and tax forms make your head spin, you're not doing your business any favors by muddling through them alone. Instead, consider hiring an accountant who can ensure everything is handled correctly, freeing you up to focus on what you do best.

2. THE VALUE OF HIRING EXPERTS

Hiring professionals to fill in the gaps doesn't just save you time—it adds real value to your business. Experts bring specialized knowledge and experience that you simply can't match by trying to do it all yourself. They can spot issues you might miss, offer solutions you hadn't considered, and help you avoid costly mistakes.

For instance, a skilled marketer can craft campaigns that reach your target audience more effectively than you could on your own. A seasoned HR consultant can help you navigate complex employment laws and build a strong team. A dedicated IT professional can ensure your systems are secure and efficient, preventing the kind of technical headaches that could disrupt your operations.

3. FOCUS ON WHAT YOU DO BEST

As a founder, your time and energy are your most valuable resources. The more you spread yourself thin trying to manage every aspect of your business, the less time you have to focus on the areas where you can make the biggest impact. By delegating tasks to professionals, you can concentrate on your core competencies—the things that inspired you to start your business in the first place.

This isn't about shirking responsibility; it's about being strategic. Successful entrepreneurs know that their role is to steer the ship, not to be down in the engine room doing everything themselves. By focusing on what you do best and letting others handle the rest, you position your business for growth and success.

4. THE COST OF DOING IT ALL YOURSELF

Some founders hesitate to hire professionals because they see it as an unnecessary expense. However, the cost of trying to do it all yourself can be much higher. Mistakes made in areas where you lack expertise can be expensive to fix and can set your business back significantly.

For example, a botched tax return can lead to fines and penalties, while poor marketing decisions can result in wasted budgets and missed opportunities. By investing in professionals from the start, you're actually protecting your business from these kinds of setbacks and ensuring a smoother path to success.

5. BUILDING A STRONG TEAM

Hiring professionals isn't just about filling in gaps—it's about building a strong, well-rounded team that can support your business as it grows. Each team member brings their own expertise, which collectively strengthens your business and allows it to operate more effectively.

When you surround yourself with capable people who complement your skills, you create a more dynamic, resilient business. You can tackle challenges more efficiently, innovate more effectively, and scale more confidently. Plus, knowing you have a solid team behind you can give you the peace of mind to take on bigger risks and pursue larger opportunities.

6. THE BOTTOM LINE: DON'T TRY TO DO IT ALL

In the early stages of your business, it's natural to want to keep costs down by doing as much as possible yourself. Yet as your business grows, you'll quickly realize that you can't—and shouldn't—do it all. By recognizing your skills gaps, hiring the right professionals, and focusing on what you do best, you'll be setting your business up for long-term success.

Remember, you don't have to be the jack of all trades to be a successful entrepreneur. You just need to be smart about where you invest your time and who you bring on board to help you achieve your vision.

LESSON 28: WORK-LIFE BALANCE IS OVERRATED

Let's talk about something that's often portrayed as the holy grail for entrepreneurs: work-life balance. We've all heard about it, read about it, and maybe even tried to achieve it. Yet, here's the thing—work-life balance is often more myth than reality, especially for those of us in the trenches of building a business.

1. THE MYTH OF PERFECT BALANCE

The idea of achieving a perfect work-life balance can stir up a lot of unnecessary guilt and stress. You might feel like you're failing if you're not able to juggle work, family, friends, and self-care perfectly. The reality is that running a business is demanding, and there will be times when work has to be prioritized. That's not failure; it's reality.

As an entrepreneur, there are seasons when you'll be all in—working late nights and weekends and sacrificing personal time to meet deadlines, close deals, or push through a crucial phase of growth. That's okay. The key is to recognize that this is part of the journey and not to beat yourself up over it.

2. INTEGRATION OVER BALANCE

Instead of striving for a perfect balance, focus on integrating work and life in a way that aligns with your priorities and values. It's about finding a rhythm that works for you, rather than forcing yourself into a rigid structure that might not be realistic. Sometimes, work will take center stage and other times; you'll be able to step back and give more attention to your personal life.

The goal isn't to separate work and life into neat, equal compartments—it's to blend them in a way that feels right for you. This might mean taking a call while on a family vacation or working on a project late at night after the kids are in bed. It's about making choices that allow you to be present in both your professional and personal life, even if that means those lines blur from time to time.

3. EMBRACE FLEXIBILITY

Flexibility is your best friend when it comes to managing the demands of entrepreneurship. The traditional 9-to-5 schedule doesn't apply to most entrepreneurs, so why should the concept of work-life balance? Instead of trying to stick to a fixed schedule, embrace the flexibility that comes with being your own boss.

This might mean working in bursts—putting in intense hours during a launch, followed by a slower period where you can recharge and focus on personal matters. It could mean setting your own hours, so you can be there for important family events or take time off when you need it. Flexibility allows you to adjust to the ebb and flow of both work and life without the pressure to maintain a perfect balance at all times.

4. PRIORITIZE WHAT MATTERS MOST

The truth is, you can't do it all—at least not all at once. So, it's essential to prioritize what matters most to you. That might mean missing a few social events to focus on a big project or scaling back on work to spend time with loved ones during important moments. It's about making conscious choices that reflect your values and understanding that it's okay to let some things slide in favor of what's truly important.

One strategy is to identify the non-negotiables in your life—those things that you won't compromise on, whether it's family time, health, or a particular work goal. Focus on these and let the rest find its place around them.

5. THE GUILT TRAP

Guilt often accompanies the pursuit of work-life balance, especially when you feel like you're falling short in one area or another. However, it's crucial to remember that balance doesn't mean giving equal time and energy to everything. It means being intentional with your time and accepting that there will be trade-offs.

Don't let guilt dictate your decisions. Instead, trust that you're making the best choices for your situation. Remember, every entrepreneur's journey is different, and what works for someone else might not work for you.

6. THE BOTTOM LINE: FIND YOUR OWN RHYTHM

Work-life balance is a nice idea, but it's not always practical for entrepreneurs. Instead of chasing an elusive ideal, focus on finding a rhythm that works for you. Embrace flexibility, prioritize what truly matters, and let go of the guilt. Your journey is unique, and it's up to you to define what balance—or integration—looks like in your life.

Remember, it's not about perfectly dividing your time between work and life—it's about making choices that allow you to thrive in both.

PART 5: INNOVATION AND ADAPTABILITY

LESSON 29: COPY, PASTE, INNOVATE

"Copy, paste, innovate." It's a phrase I first heard from Fadi Ghandour, a Lebanese-Jordanian entrepreneur who has built a legacy of success and credibility. Ghandour is the type of entrepreneur we should all be learning from—the real deal, not one of those "fake it until you make it" content creators that flood social media. His words struck a chord with me, especially in the world of marketing and content creation, where innovation is often just a smarter version of something that already exists.

In most cases, you don't need to reinvent the wheel to achieve success. Sometimes, the best approach is to find a concept that's already working, see how you can make it better, and innovate from there. It's a method that has propelled countless startups to success.

1. DON'T REINVENT THE WHEEL—PERFECT IT

Take a look at some of the most successful companies today, and you'll see that many of them didn't start with a groundbreaking, original idea. Instead, they took an existing concept, copied it, and then tailored it to fit a specific market or made it better. Careem, for example, took Uber's model and innovated it for the Middle Eastern market, addressing local needs and preferences. Airbnb did something similar by taking the concept of hotel listings and expanding it to include apartments, fundamentally changing how people think about travel accommodations.

The lesson here is clear: find a winning idea, understand it deeply, and then see how you can improve or customize it for a particular audience or region. This approach is not only efficient but also reduces the risk of venturing into completely uncharted territory.

2. LOOK FOR INSPIRATION EVERYWHERE

Sometimes, the best ideas come to you when you're least expecting them. Imagine you're traveling in another country and come across a business concept that's incredibly smart but doesn't exist in your home market. Ask yourself, "Is this something I can copy, paste, and innovate back home?" If the answer is yes, you may have just found a winning business idea.

The beauty of this approach is that it doesn't require you to come up with something entirely new. Instead, it's about recognizing good ideas when you see them and figuring out how to make them even better. It's about understanding that innovation doesn't always start from scratch—it often begins with a foundation that someone else has already laid.

3. CONTENT CURATION IN MARKETING

In the marketing world, this approach is particularly valuable. My team often comes across brilliant campaigns, clever content ideas, or eye-catching designs. Rather than starting from zero every time, we practice what I call content curation, not creation. We take that smart idea, copy the essence of it, modify it to fit our brand, and create our own version.

This isn't about plagiarism or lack of originality. It's about being resourceful and recognizing that the creative process is often about building on what's already out there. By curating and refining existing ideas, you can produce something that's both innovative and uniquely yours.

4. INNOVATE WITHOUT HESITATION

So, the next time you come across a successful idea, whether in your industry or elsewhere, don't be afraid to ask yourself how you can make it better. Copy the core concept, paste it into your context, and innovate from there. The path to success doesn't always require you to blaze a completely new trail—sometimes, it's about perfecting the path that's already been laid.

Remember, innovation is often about taking what works and making it work even better. So don't hesitate to copy, paste, and innovate your way to success.

LESSON 30: TURNING CHALLENGES INTO OPPORTUNITIES

In 2024, the pace of change is faster than ever, and as an entrepreneur, you must adapt quickly. This lesson is about not fearing change but embracing it, even loving it. Change brings new opportunities, and rather than seeing it as a threat, you should see it as a gateway to growth.

No matter what field your business operates in, everything is evolving. Whether it's due to advancements in algorithms, machine learning, artificial intelligence, or other emerging technologies, we're witnessing a growth pace higher than at any previous time, and this trend is only going to accelerate in the coming years.

1. THE POWER OF ADAPTATION

Back in 2021, when we founded OBCIDO, a marketing agency, we offered services similar to those provided by other agencies. We were doing well, attracting clients and establishing our market share. As the CEO, I wasn't content with just keeping up—I wanted to lead. I quickly learned that change isn't something to fear; it's something to embrace.

I made it a priority to adopt every new technology that could push our business forward. Whether it was offering marketing services for NFT projects or utilizing AI to boost productivity, we were always looking for ways to stay ahead. Here's an example: we started as a company offering web development solutions with a team of full-stack developers who could build beautiful websites, then we encountered a new challenge: what the heck is smart contract development?

2. THE SMART CONTRACT OPPORTUNITY

Let me explain briefly: a smart contract is a self-executing contract with the terms of the agreement directly written into code. While a website might sell for an average of $5,000, a smart contract could sell for $10,000 to $20,000 or even more. The development costs for the agency? Almost the same. However, there were far fewer agencies offering smart contract development compared to web development, and we became one of the few that excelled in both.

Despite our foray into smart contract development, we didn't abandon web development. This is a crucial point: you should never fully invest all your resources in a trend, but you should capitalize on it while it lasts. We became one of the few agencies that could excel in Web 3.0 development while still providing top-notch marketing for all types of businesses.

3. SEIZING OPPORTUNITIES, NOT FEARING THEM

While others might have insisted on sticking strictly to traditional marketing services and been wary of Web 3.0, we saw the change as an opportunity. We capitalized on the hype, tripling our revenue in some months, all while keeping our core foundation strong. This is just one example of how we've embraced change repeatedly, and it's a strategy we continue to follow.

Now, when we see change, we don't see a threat—we see opportunity. It all comes down to mindset. You can choose to learn, grow, and evolve, or you can choose to be fearful and miss out. The choice is yours.

4. CONTINUOUS LEARNING AND ADAPTATION

Embracing change means being ready to learn fast, especially as new technologies emerge. As an entrepreneur, you can't afford to be complacent—you need to be the first to learn and adapt. That's why, even in your free time, it's essential to prioritize your resources. Everything you need to learn is out there and often free. Follow the right accounts on social media, watch educational YouTube videos, and take online courses—there are endless opportunities to grow.

In the end, embracing change isn't just about surviving in a rapidly evolving world—it's about thriving. It's about turning every challenge into an opportunity and every new development into a chance to grow. So, the next time you're faced with change, don't hesitate. Embrace it, learn from it, and use it to propel your business forward.

LESSON 31: PLANS ARE JUST ASSUMPTIONS

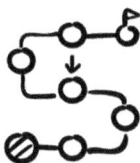

If planning always worked, business would be a risk-free endeavor. Yet, as every entrepreneur quickly learns, that's far from reality. Planning is important—it gives us direction, a sense of purpose, and a roadmap to follow. Yet, the world we operate in is far too complex to be neatly packaged into a long-term plan. There are simply too many variables beyond our control: market conditions, competitors' moves, customer behaviors, economic shifts—the list goes on. Writing a plan might provide a comforting sense of control, but in many cases, that control is just an illusion.

As a CEO and founder, I've come to see business plans as educated guesses rather than definitive guides. Marketing plans? Guesses. Strategic plans? Guesses. This mindset shift has allowed me to prioritize what truly matters—action and adaptability—without being shackled by the anxiety of trying to predict the unpredictable.

1. THE TRAP OF RIGID PLANNING

When we start to confuse our guesses with guarantees, we fall into the trap of rigid planning. Strict plans can lock us into a path that may have made sense at the time of conception but no longer aligns with reality. The problem arises when we stick to these plans simply because we are committed to them, even when new information suggests a better route. In doing so, we risk becoming inflexible, unable to pivot when necessary.

In the dynamic environment of a startup, flexibility is not just a nice-to-have; it's essential. I've learned that as opportunities or challenges arise, I need the freedom to pivot quickly, saying, "This is the right move today." Timing is everything, and more often than not, long-term plans miss the mark because they are made with the least amount of actionable information. The best insights come from being in the thick of the action, not from forecasting months or years ahead.

2. THE REALITY OF PLANS IN PRACTICE

This doesn't mean I ignore the future altogether. I think about the future, but I've stopped obsessing over detailed, long-term plans. In my experience, these plans often end up gathering dust, forgotten in the face of reality. Instead, I focus on what needs to happen this week, not what might happen this year. I make decisions when I have the most information, not when I'm least informed.

It's okay to wing it. Sometimes, just taking the first step is enough —you can figure out the details as you go. Working without a strict plan might seem risky, but following a rigid plan that doesn't match reality is even riskier.

3. REAL-WORLD EXAMPLES: WHEN PLANNING FAILS

This isn't to say that planning has no place in business. It does— but it's crucial to recognize its limitations. A study by PricewaterhouseCoopers, which analyzed 10,640 projects across 200 companies in 30 countries, revealed that only 2.5% of these companies successfully completed 100% of their projects. This statistic is telling that even with detailed plans and considerable resources, most projects don't go as expected. Plans often crumble under the weight of unforeseen challenges and shifting circumstances.

Marketing, in particular, is an area where rigid planning can backfire spectacularly. Take the case of Gap's 2010 logo redesign. The company spent a staggering $100 million on rebranding, confident that their new logo and marketing push would resonate with consumers. Instead, they faced immediate backlash. Within 24 hours, thousands of negative comments flooded social media, and parody logos went viral. Just a week later, Gap reverted to its original logo, abandoning the new design altogether. This high-profile failure underscores a critical lesson: even the most meticulously planned and well-funded strategies can fall flat if they don't align with current market realities.

These examples remind us that no matter how much time and money we invest in planning, success is never guaranteed. The market is unpredictable, and the only certainty is that things will change.

4. EMBRACING FLEXIBILITY AND IMPROVISATION

So, what's the alternative? It's about striking a balance between planning and flexibility. I'm not suggesting you throw planning out the window—far from it. Planning is essential for setting goals, allocating resources, and giving your team direction. However, it's equally important to remain flexible, and to be ready to pivot when new information comes to light or when the market takes an unexpected turn.

In my experience, the most effective strategy is to focus on short-term objectives that can be adjusted as needed. I concentrate on what needs to happen this week, not what might happen next year. This approach allows me to make decisions based on the best information available at the moment, rather than relying on predictions made months in advance.

Flexibility also means being willing to improvise. Some of the best decisions we've made at OBCIDO were not part of any plan. They were responses to unforeseen opportunities or challenges—decisions made at the moment, with the full understanding that we could adapt as we went along.

5. THE COURAGE TO WING IT

Embracing this mindset requires a certain level of courage. It's not easy to let go of the comfort that comes with a detailed plan. In reality, working without a strict plan might seem risky, but following a rigid plan that no longer aligns with reality is even riskier.

It's okay to wing it, take that first step, and figure out the details as you go. The key is to stay nimble, to keep your eyes and ears open, and to be ready to change course when necessary. Planning is important, but it's just the starting point. The real work begins when you're in the trenches, dealing with the day-to-day realities of running a business.

In the end, remember that plans are just assumptions, and the ability to adapt is what will set you apart. By staying flexible and open to improvisation, you'll be better equipped to navigate the unpredictable world of entrepreneurship and lead your business to success.

LESSON 32: BETTER TO ACT AND FAIL THAN DO NOTHING

Let's be real—how many times have you found yourself stuck in the endless loop of "What if?" You know the one: *What if I make the wrong decision? What if this doesn't work out? What if I fail?* It's the kind of thinking that can keep you up at night, staring at the ceiling, paralyzed by the fear of making a move. Here's a little secret: doing nothing is often worse than doing something and failing. Yep, you heard that right. It's far better to take a shot and miss than to sit on the sidelines, wondering what could've been.

Think of it this way: inaction is like hitting the pause button on your progress. You're not moving forward, you're not learning, and worst of all, you're not seizing opportunities that could take your business to the next level. Sure, taking action comes with risks, but so does doing nothing. In fact, inaction is its own kind of risk—a silent one that slowly but surely eats away at your potential.

So, what's the solution? Simple: take the leap. Act. Even if it means you might stumble, even if you're not 100% sure of the outcome, just go for it. Because the truth is, you'll learn a heck of a lot more from a failed attempt than from doing nothing at all.

1. THE PARALYSIS OF INACTION

One of the biggest challenges many entrepreneurs face is the fear of making mistakes. This fear can lead to analysis paralysis, where you spend so much time weighing options and considering potential outcomes that you end up taking no action at all. The problem with this approach is that while you're stuck in indecision, opportunities are passing you by.

In business, timing is everything. Waiting too long to make a move can mean missing out on key opportunities or falling behind your competitors. A report by the Harvard Business Review found that companies that move quickly to take advantage of market opportunities are 50% more likely to become market leaders than those that take a slower, more cautious approach.

2. THE VALUE OF FAILURE

It's important to recognize that failure isn't the opposite of success; it's a part of the journey toward it. Every successful entrepreneur has faced setbacks and made mistakes along the way. What sets them apart is their willingness to learn from those failures and keep moving forward.

When you take action, even if it leads to failure, you gain valuable insights that you wouldn't have otherwise. You learn what works, what doesn't, and how to improve for the future. Each failure brings you one step closer to success.

Thomas Edison, one of the most prolific inventors in history, famously said, "I have not failed. I've just found 10,000 ways that won't work." Edison's relentless pursuit of innovation, despite countless failures, ultimately led to some of the most significant technological advancements of his time. His story is a powerful reminder that failure is not something to be feared but embraced as a crucial part of the learning process.

3. TAKING CALCULATED RISKS

This isn't to say that you should act recklessly or make decisions without proper consideration. The key is to take calculated risks—actions that are informed by research, experience, and intuition, but not paralyzed by the fear of failure.

When you take a calculated risk, you accept that there's a possibility of failure, but you also recognize that doing nothing is often the greater risk. Inaction guarantees that nothing will change, while taking a chance opens the door to new opportunities, growth, and success.

4. THE COST OF DOING NOTHING

Doing nothing isn't just about missed opportunities; it can also have a tangible cost. In the business world, standing still often means falling behind. Markets evolve, customer preferences shift, and competitors innovate. If you're not moving forward, you're effectively moving backward.

Consider the rise of digital marketing. Companies that hesitated to embrace digital strategies early on found themselves struggling to catch up as the market shifted online. Meanwhile, businesses that took the leap—experimenting with social media, content marketing, and e-commerce—reaped the rewards of early adoption, even if some of their initial efforts weren't perfect.

Another example is Blockbuster, which famously passed up the opportunity to buy Netflix for $50 million in 2000. At the time, Blockbuster was the dominant player in the video rental industry, and Netflix was a small, unproven startup. Blockbuster's decision to do nothing, to stick with its traditional business model rather than adapt to the changing landscape, ultimately led to its downfall. Meanwhile, Netflix, now valued at over $200 billion, became a global powerhouse by embracing change and taking calculated risks.

5. EMBRACE A BIAS FOR ACTION

One of the most valuable traits you can cultivate as an entrepreneur is a bias for action. This means developing a mindset that prioritizes action over endless deliberation. It's about recognizing when you have enough information to make a decision and then taking that leap of faith.

A bias for action doesn't mean acting without thought—it means not letting fear, uncertainty, or the pursuit of perfection hold you back. It means understanding that progress often comes from making mistakes, learning from them, and iterating quickly.

Amazon's founder, Jeff Bezos, is a strong advocate of this mindset. He coined the term "Day 1 mentality," which refers to the idea that every day is a new opportunity to take action, innovate, and disrupt the status quo. Bezos emphasizes that in a fast-moving world, the ability to make quick decisions and act on them is what keeps a company agile and competitive.

6. ACTION CREATES MOMENTUM

When you take action, you create momentum. One small step leads to another, and before you know it, you're making significant progress. This momentum is critical in business, where the ability to move quickly and decisively can be the difference between success and failure.

Even if your first step isn't perfect, it's better than standing still. You can always course-correct along the way, but you can't steer a ship that's anchored in place. The simple act of moving forward—of taking that first step—sets things in motion and opens up new possibilities.

7. THE BOTTOM LINE: DON'T LET FEAR HOLD YOU BACK

In the end, the biggest risk you can take as an entrepreneur is to do nothing. Fear of failure, fear of making the wrong decision, or fear of the unknown can paralyze you and prevent you from reaching your full potential. Remember, the most successful entrepreneurs aren't those who avoid failure—they're the ones who aren't afraid to fail, learn, and keep going.

So, the next time you're faced with a decision, don't let fear or uncertainty hold you back. Take action, embrace the possibility of failure, and trust that even if things don't go as planned, you'll come out stronger and wiser on the other side. After all, it's better to act and fail than to do nothing and regret the opportunities you missed.

LESSON 33: BUSINESS IS A SKILL, NOT A SCIENCE

Imagine trying to learn how to drive a car by reading a book. You might memorize every road sign, understand the mechanics, and even ace a written exam. Yet, when you finally get behind the wheel, it's a whole different story. The same goes for business. You can study all the theories, read every case study, and earn multiple degrees, but none of that will make you a skilled entrepreneur. Business, like driving, is something you learn by doing.

1. THE MYTH OF THE PERFECT FORMULA

In school, we're often taught that there's a right way to do everything—a formula for success. However in the real world, especially in business, there is no one-size-fits-all approach. What works for one company might not work for another, and what brings success today might be irrelevant tomorrow. Business isn't a science with fixed rules and predictable outcomes; it's a skill that you develop through experience, trial and error, and a willingness to learn.

You can spend years studying business models, market strategies, and financial forecasts, but none of that guarantees success. In fact, the real learning begins when you take the plunge, make mistakes, and figure out how to navigate the unpredictable terrain of the business world.

2. THE POWER OF DOING

The best way to learn business is by doing. It's about getting your hands dirty, making decisions, facing challenges, and learning from your experiences. Every successful entrepreneur will tell you that their most valuable lessons didn't come from textbooks or lectures —they came from real-world experiences, both good and bad.

When you step into the world of business, you're going to make mistakes. You'll encounter setbacks, face uncertainty, and experience failures. Still, it's through these experiences that you'll develop the skills, instincts, and resilience needed to succeed. As you navigate the ups and downs, you'll start to see patterns, recognize opportunities, and build the confidence to take calculated risks.

3. FEARLESSNESS IS A MUSCLE

Arianna Huffington, the founder of The Huffington Post, once said, "Fearlessness is like a muscle. I know from my own life that the more I exercise it, the more natural it becomes to not let my fears run me." This couldn't be truer in business. The more you step out of your comfort zone and face your fears, the more skilled you become at managing them.

In business, fear can be a powerful inhibitor. It can stop you from taking risks, pursuing opportunities, or making bold decisions. The only way to overcome fear is to confront it head-on. The more you do, the more natural it becomes to act in the face of uncertainty.

Over time, you'll find that what once seemed terrifying is now just another challenge to tackle.

4. EMBRACE THE LEARNING PROCESS

Learning business is not a linear process. It's messy, unpredictable, and full of surprises. That's what makes it exciting. Every challenge you face is an opportunity to grow, every mistake is a lesson in disguise, and every success is a testament to your hard-earned skills.

The key is to embrace the process. Don't be afraid to make mistakes, and don't be discouraged by setbacks. Instead, view each experience as a valuable lesson that's helping you become a better entrepreneur. Remember, business is not about following a predetermined path—it's about forging your own.

5. CONTINUOUS IMPROVEMENT

Just like any skill, business requires continuous practice and improvement. The market is constantly changing, new technologies are emerging, and customer expectations are evolving. To stay ahead, you need to keep learning, adapting, and refining your approach.

This doesn't mean you need to go back to school or earn another degree. It means staying curious, seeking out new experiences, and being open to change. Whether it's trying out a new strategy, experimenting with a different business model, or learning a new tool, continuous improvement is key to staying competitive.

6. THE BOTTOM LINE: GET BEHIND THE WHEEL

You wouldn't expect to become a skilled driver by reading a manual, so don't expect to become a successful entrepreneur by only studying business theories. Business is a skill—one that you can only master by getting behind the wheel and taking it for a spin.

It's about taking risks, making decisions, learning from your experiences, and building your confidence along the way. The road will be full of twists and turns, but with each mile, you'll become more skilled, more resilient, and more fearless. So, get out there, start doing, and remember: business isn't a science—it's a skill, and the best way to learn it is by living it.

LESSON 34: IDEAS ARE FREE, LAUNCHES AREN'T

We all have those moments when a brilliant idea strikes us out of nowhere. It feels exhilarating, like you've just unlocked the key to success. Here's the thing: ideas are free. They come and go, and everyone has them. What really matters is the execution—turning that idea into reality. That's where things get a little more complicated and, yes, expensive.

When you're buzzing with excitement over a new idea, it's easy to get carried away. You start imagining all the possibilities, the potential profits, the impact it could have. Yet, before you dive in headfirst, there's a crucial question you need to ask yourself: Can I afford to launch this?

It's one thing to have a groundbreaking idea, but it's another to bring it to life within the constraints of your budget. You might dream of launching the next big tech startup, producing a line of luxury products, or even exploring more ambitious ventures like oil production or satellite launches. These aren't just ideas—they're massive undertakings that require equally massive investments.

1. THE REALITY CHECK

Let's be real: not every idea is meant to be pursued, at least not right away. The cost of launching an idea can vary greatly depending on the industry, the scale, and the resources required. It's easy to get swept up in the excitement and overlook the practicalities, but this is where many entrepreneurs stumble. You need to fit the launch of your idea within the size of your own wallet.

Take a step back and evaluate your financial situation. Do you have the resources to launch this idea without stretching yourself too thin? Are you prepared for the unexpected costs that often arise during the launch process? It's important to be honest with yourself about what you can realistically afford.

2. STARTING SMALL, THINKING BIG

Just because you can't afford a large-scale launch right now doesn't mean you should abandon your idea altogether. Sometimes, the best approach is to start small and scale up over time. Test the waters with a smaller version of your idea—a pilot project or a minimum viable product (MVP). This allows you to gauge interest, gather feedback, and refine your concept without breaking the bank.

By starting small, you can also prove the viability of your idea to potential investors or partners down the line. Once you've demonstrated that there's a market for your product or service, it becomes easier to secure the funding needed for a larger-scale launch.

3. THE IMPORTANCE OF BUDGETING

Budgeting isn't just about keeping track of expenses; it's about making strategic decisions that ensure your idea has the best chance of success. When planning your launch, be meticulous about your budget. Include every possible cost—production, marketing, distribution, legal fees, and any other expenses that might arise. Then, add a buffer for the unexpected, because things rarely go exactly as planned.

If you find that your launch budget is exceeding what you can afford, it's time to reconsider your approach. Look for areas where you can cut costs or explore alternative funding options like crowdfunding, grants, or partnerships. Whatever you do, don't stretch yourself so thin that one setback could derail your entire project.

4. KNOW WHEN TO PIVOT

Sometimes, despite your best efforts, you might realize that your idea is too costly to launch as originally planned. This doesn't mean you have to give up—it might just mean you need to pivot. Look for ways to adapt your idea to fit within your budget. Can you simplify the concept? Target a smaller market? Use different materials or methods? Being flexible and open to change can save you from overextending yourself financially.

Remember, the goal is to get your idea off the ground, not to drain your resources in the process. By staying within your budget, you give yourself the best chance of success without unnecessary risk.

5. THE BOTTOM LINE: BE REALISTIC, BE STRATEGIC

Ideas are exciting, and the possibilities they represent are endless. Launching an idea, however, is a whole different ball game. It requires careful planning, realistic budgeting, and sometimes, a little bit of restraint. The best ideas are the ones that are not only creative and innovative but also feasible.

So, before you jump into your next big idea, make sure it fits within the size of your wallet. Be strategic, be realistic, and remember that sometimes, starting small can lead to big things.

LESSON 35: FACE IT TO FIX IT

Every entrepreneur encounters problems—it's an inevitable part of running a business. The way you handle these problems can make all the difference between overcoming challenges and getting stuck in them. The first and most crucial step to solving any problem is simple: admit it exists.

1. THE POWER OF ACKNOWLEDGMENT

It's easy to ignore problems or downplay them, hoping they'll resolve on their own. The truth is that problems rarely disappear without action. To fix something, you first have to face it head-on. Acknowledge that there's an issue, even if it's uncomfortable or daunting. This act of admission is empowering because it shifts you from a state of denial to one of action.

By admitting the problem, you're already halfway to finding a solution. It's like shining a light in a dark room—once you see what's there, you can start to navigate your way through it.

2. WRITE IT DOWN

One effective strategy to confront and understand a problem is to write it down. There's something about putting pen to paper that clarifies your thoughts and helps you see the issue more objectively. Writing it down forces you to articulate exactly what the problem is, which can reveal aspects you hadn't considered before.

As you write, try to pinpoint the root cause of the problem. Ask yourself: What led to this situation? What are the underlying issues? By breaking it down into smaller parts, you'll gain a clearer understanding of the problem, making it easier to address.

3. STEP BACK AND GET PERSPECTIVE

Once you've acknowledged and written down the problem, it's time to take a step back. Often, when we're too close to a situation, our perspective becomes narrow, and we struggle to see the bigger picture. By distancing yourself, even just mentally, you can gain a new angle on the problem.

Consider looking at the issue from different perspectives—how would your customers view it? What about your employees or stakeholders? Sometimes, asking for input from others can provide insights you hadn't considered. A fresh perspective can open up new avenues for solutions that weren't apparent before.

4. ANALYZE BEFORE ACTING

It's tempting to jump straight into problem-solving mode, but taking the time to analyze the situation first is key to finding the right solution. Rushing to fix a problem without fully understanding it can lead to quick fixes that don't address the root cause—and the problem may resurface later.

Give yourself time to analyze the situation thoroughly. Consider all possible causes and potential impacts of the problem. The more you understand about the problem itself, the more effective your solution will be.

5. TURN PROBLEMS INTO OPPORTUNITIES

Sometimes, problems can be opportunities in disguise. By facing and fixing a problem, you might uncover a chance to improve your business, streamline a process, or innovate in a way that sets you apart from the competition. Approach problems with a mindset that looks for the silver lining—what can you learn from this? How can this challenge push your business forward?

6. THE BOTTOM LINE: FACE IT, FIX IT, AND MOVE FORWARD

Problems are a natural part of business, but ignoring them or pretending they don't exist won't make them go away. By admitting the problem, writing it down, gaining perspective, and taking the time to analyze it, you position yourself to not only solve the issue but to learn and grow from it.

Remember, the faster you face a problem, the quicker you can fix it and move forward. Don't let fear or denial hold you back— embrace challenges as opportunities to make your business stronger and more resilient.

PART 6: CONTINUOUS LEARNING AND DEVELOPMENT

LESSON 36: HOW TO GET YOUR COLD EMAILS READ

As a CEO, I receive countless cold emails, LinkedIn messages, and other outreach attempts every day. They flood my inbox, my DMs, and my notifications. Here's the thing—only a few of them actually catch my eye. The rest? They're skimmed or skipped entirely.

Why do some messages stand out while others fade into the background? It's simple: the successful ones are short, clear, and straight to the point.

When you're reaching out—whether it's through email, LinkedIn, or any other platform—remember that your prospect doesn't know you yet. That means they don't care about your company's story, your accolades, or your expertise—at least not at first. What they care about is whether you can solve their problem. You've only got a few seconds to convince them that you can.

Here's what I've learned from both sending and receiving cold outreach:

1. MAKE YOUR PROSPECT THE HERO

Your outreach isn't about you—it's about them. Don't lead with your achievements or your company's success. Instead, focus on your prospect's business, their pain points, and how you can add value. They need to see themselves as the hero of the story, with you as the guide who can help them achieve their goals.

2. BE CLEAR AND DIRECT

Clarity is key. In a world full of distractions, people don't have the time or patience to wade through long, meandering messages. Get to the point quickly. What do you do? How can you help? Why should they care? If you can answer these questions within the first few lines, you've already set yourself apart from the majority of cold outreach.

3. SHOW YOUR VALUE UPFRONT

The best cold emails I've received all had one thing in common: they demonstrated value right from the start. Whether it was a case study, a specific solution to a known problem, or a unique insight into my business, these messages made it clear that the sender had done their homework and had something valuable to offer. On the flip side, long, vague emails that beat around the bush? They get skipped.

4. KEEP IT SHORT

You don't need to tell your whole story in one message. In fact, the shorter, the better. A few well-chosen sentences can be far more effective than a lengthy essay. Respect your prospect's time by delivering your message in a concise, easy-to-digest format.

5. BE STRAIGHTFORWARD AND TRANSPARENT

There's no need for gimmicks or overly clever language. Be straightforward about what you're offering and why it matters. Transparency builds trust, and trust is essential in any business relationship—especially when you're reaching out cold.

6. ADD VALUE TO EVERY INTERACTION

Every interaction with a prospect should add value, whether it's the first email or a follow-up message. Think of each touchpoint as an opportunity to build trust and demonstrate your expertise. Over time, these small, value-driven interactions can lead to meaningful business relationships.

Cold outreach is a crucial part of growing a startup, but it's also one of the most challenging. To stand out in a sea of messages, you need to focus on your prospect's needs, communicate your value clearly, and make every word count. When you make your prospect the hero of the story, you'll find that your messages start to get the attention they deserve.

LESSON 37: THE ART OF THE FOLLOW-UP

In the world of sales, follow-ups can be the difference between closing a deal and losing a lead. However, there's a fine line between being persistent and being pushy. When you have a high-quality service to offer, it's crucial to keep your value high without crossing into the territory of being overbearing.

After your initial discovery call, a well-timed follow-up can reinforce your value and keep the conversation going. Yet, how many follow-ups are too many? In my experience, the sweet spot is one or two. Beyond that, you risk creating a negative impression, even if your offer is top-notch.

1. QUALITY OVER QUANTITY

One or two well-crafted follow-ups can be far more effective than a barrage of messages. High-pressure tactics often do more harm than good, making potential clients wary and turning them off your offer. Instead, focus on maintaining high sales standards. A concise, value-driven follow-up shows that you respect your prospect's time and decisions, reinforcing your professionalism.

2. OFFER VALUE UPFRONT

Your follow-up isn't just a reminder—it's another opportunity to offer value. Share a piece of useful information, a relevant case study, or a unique insight that addresses their pain points. This not only keeps the conversation relevant but also positions you as a trusted resource rather than just another salesperson.

3. KNOW WHEN TO MOVE ON

Not every lead will convert, and that's okay. Don't get stuck chasing a single prospect to the point where you start to devalue your own offer. If you've followed up once or twice without a response, it's time to move on. Respect your time and theirs, and leave the door open for future opportunities without being pushy.

4. LET YOUR FIRST IMPRESSION DO THE WORK

I've had leads come back months later, not because I hounded them with follow-ups, but because I made a great, generous first impression. When you provide value upfront and show that you're confident in what you offer, it leaves a lasting impression. Sometimes, all it takes is a bit of time for the prospect to realize the true value of what you bring to the table.

5. SHARE FREELY, LET THINGS FLOW NATURALLY

Don't hold back valuable information during your initial contact. Share it freely, and let the relationship develop naturally. A respectful, value-driven follow-up can make all the difference in how you're perceived. It shows that you're confident in your offer and that you're focused on the prospect's needs, not just your own sales targets.

In the end, mastering the follow-up is about balance. Offer value, follow up respectfully, and then move on if the timing isn't right. When you maintain high standards and focus on the long-term relationship rather than the immediate sale, you'll find that your follow-ups become a powerful tool in your sales strategy—without ever compromising your value.

LESSON 38: THE ULTIMATE INVESTMENT

Over the years, I've learned that the best form of investment isn't found in stocks, real estate, or even the latest cryptocurrency trend —it's investing in yourself. While this might sound obvious, I've found that many people don't fully grasp what self-investment really means. Here's the thing: you don't have to be a business owner or an entrepreneur to reap the benefits of this lesson. It applies to everyone, regardless of their career path.

Let me give you an example from my industry—marketing.

Imagine you're a graphic designer making $20 an hour. You're good at what you do, but you're looking to grow, increase your value, and ultimately earn more. There are several ways you could achieve this, especially in a fast-paced environment like a startup, where growth is often rapid for those who put in the work.

1. SEEK PROMOTIONS

The first route is the most straightforward—aim for a promotion. In startups, talent doesn't go unnoticed for long. If you're consistently delivering high-quality work and taking initiative, the opportunity for a promotion is often within reach.

2. LEARN NEW SKILLS

The second route, and perhaps the most powerful, is to invest in learning new skills. Let's say that, as a graphic designer, you decide to learn UX/UI design and tools like Figma. Suddenly, you're not just a designer working on social media content—you're now capable of designing entire websites. This new skill set not only doubles your value but also opens up new opportunities for career advancement.

Startups are always on the lookout for versatile talent—people who bring more to the table than what's listed in their job description. As an entrepreneur, I'm constantly seeking individuals who have invested in themselves and have taken the initiative to expand their skill sets. These are the people who stand out, the ones who get noticed, and the ones who grow rapidly within the company.

3. THE PITFALLS OF CHASING HIGH RETURNS

We've all heard stories of people who invested in Bitcoin, NFTs, or stocks and made fortunes overnight, but what we don't hear as often are the stories of those who lost everything chasing these "too good to be true" opportunities. Trust me, there are plenty of those stories out there.

I'm not just speaking theoretically here—I've tried both routes. I've put money into high-risk, high-reward investments, and more often than not, I ended up losing. On the other hand every new skill I've learned, and every new tool I've mastered, has paid off in ways I couldn't have imagined. The return on investment from self-education has been immeasurable.

4. KEEP LEARNING, KEEP GROWING

So, my advice is simple: don't chase the next big investment opportunity that promises overnight wealth. Instead, invest in yourself. You'd be amazed at what just three months of learning a new skill can do for your career. Now, imagine what six months, nine months, or a year could do. The possibilities are endless.

Don't limit yourself to one set of skills. Keep learning and keep growing, whether you're a founder, an executive, or a hardworking employee. The time and effort you invest in yourself will always pay dividends—more than any get-rich-quick scheme ever could.

In the end, the most secure and rewarding investment you can make is in your own growth and development. It's an investment that never loses value, and it's one that you have full control over.

LESSON 39: OUTWORKING EVERYONE IN THE ROOM

Starting a business demands two primary investments: your money and your time. Ideally, you'd invest both, as combining financial resources with dedicated effort allows you to build a strong team and achieve more faster. However, for many, especially in the early stages, one of these investments might be more accessible than the other.

1. THE LUXURY OF MONEY VS. THE POWER OF TIME

Investing money into a new business is often seen as a luxury. Some founders choose this route, pouring capital into their ventures while delegating the hard work to others. Here's the truth: while money can jumpstart your business, it's time that builds it. For most startups, especially bootstrapped ones, time is the strongest asset at your disposal.

Consider this: an agency—whether it's in marketing, SEO, web development, or any other field—can be started with almost no money but requires a significant investment of time. If you have time, use it wisely because it's your most valuable resource.

2. THE REALITY OF TIME INVESTMENT

When we bootstrapped OBCIDO and other ventures, we did so with minimal financial expenses but a tremendous investment of time. During the first year of OBCIDO, I wasn't just working hard —I was living the business. I worked and slept in the same office, putting in 14+ hour days, every day, from 8 a.m. until 1 a.m. This wasn't just a phase; it was a necessity. Even years later, I maintained a full-time position within my startup, fully immersed in its day-to-day operations.

The entrepreneurial journey is rewarding, but it's not for everyone. Not everyone can sacrifice work-life balance, and that's okay. However, if you're planning to start a business with minimal investment and without committing to long hours, you're setting yourself up for disappointment. My advice? Either rethink your plans or be prepared to roll up your sleeves and get to work.

3. THE ROLE OF THE FOUNDER: LEADING BY EXAMPLE

Here's the bold lesson: the founder must be the hardest working employee in the company. Your commitment sets the tone for the entire organization. If you're not willing to outwork everyone, you can't expect your team to do so either. The foundation of your business is built on the work you put into it, especially in those critical early stages. If that foundation is weak, the entire structure is at risk.

4. THE REALITIES OF ENTREPRENEURSHIP

The entrepreneurship route is challenging, and not everyone can make it. It's not just about the hours—it's about the relentless commitment to seeing your vision through. As a founder, your willingness to invest your time, energy, and effort sets the standard for the company's culture and its potential for success. If you want your business to thrive, you need to be its hardest-working employee. That's non-negotiable.

LESSON 40: DON'T CREATE IT, CURATE IT

One of the most valuable lessons I've learned in building a successful business is that you don't always need to create something from scratch to offer something valuable. In fact, some of the best businesses are built not on innovation but on curation.

A business, at its core, is about identifying a pain point, problem, or need, finding a solution, and offering that solution at a premium. Here's the secret: you don't have to be the one to create that solution yourself. You don't need to be a software engineer to sell software or the best designer to sell creative work. Instead, you can simply find the talent, connect with the client, offer the service, charge a premium, and pocket the profits.

This approach might seem logical, but let's take it a step further and unlock the real potential of this strategy.

1. THE POWER OF BEING THE MIDDLEMAN

When you're starting out and don't have a huge capital base, the idea of building everything yourself can be overwhelming and impractical. Whether it's a product, a service, or a SaaS solution, you can always position yourself as the middleman. Here's the secret: find an innovative SaaS solution that meets a specific need in the market. Speak to the company behind it, negotiate a white-label agreement, and make that solution your own.

Learn the software inside and out, test it rigorously, and build a compelling case study that showcases its effectiveness. Then, mark it up, promote it aggressively, and sell it as your own premium offering.

2. OFFERING VALUE THROUGH EXPERTISE

Now, you might think, "Why would a client pay me a premium when they could just buy the software directly for a fraction of the cost?" The answer is simple: they won't invest the hours that you did in training and onboarding. Your clients will gladly pay you the premium to handle all the technical work, the setup, the optimization, and the ongoing management. They're paying for your expertise, your time, and your ability to deliver results—not just for the software itself.

This same principle applies to services. Take Meta Ads Manager, for example—a tool that's available to anyone. Yet mastering it takes time, effort, and a deep understanding of digital marketing. So, what do you do? Find a service provider who can run ads effectively for a few hundred dollars a month. Study the service, learn all the ins and outs, and then market your agency as the go-to provider for this service. Find clients, sell them on your expertise, and charge thousands.

3. AVOIDING THE PITFALL OF OVER-CREATION

One of the biggest mistakes I've seen entrepreneurs make is trying to develop an entirely new SaaS solution when there are already ten other companies offering the same thing. Instead of spending years in development and hundreds of thousands of dollars, you can white-label an existing service, use it as your own, hold zero risk, and sell it at a premium.

This isn't about cutting corners—it's about being smart with your resources. It's about recognizing that the market doesn't always need another original product; sometimes, it just needs someone who can take an existing solution and make it more accessible, more user-friendly, and more effective.

4. THE FORMULA FOR SUCCESS

The formula is simple: find something that works, make it your own, and sell it better than anyone else. Curate the best solutions, add your unique value, and deliver results that exceed your clients' expectations. By doing this, you're not just building a business— you're building a reputation for being the expert who knows how to get things done.

Remember, the key to success isn't always about creating from scratch. Sometimes, the smartest move you can make is to curate what's already out there and make it work for you. Don't create it —curate it.

LESSON 41: EDUCATION DOESN'T END WITH GRADUATION

Let's get one thing straight: if you think learning stops the day you walk across that stage in your cap and gown, you're setting yourself up for a rude awakening. In the fast-paced world of entrepreneurship, the moment you stop learning is the moment you start falling behind. The most successful entrepreneurs understand that education isn't a box you check off—it's a lifelong commitment. Whether it's mastering new business practices, staying updated on industry trends, or simply expanding your general knowledge, continuous learning is key to staying ahead of the curve.

1. THE LIFELONG STUDENT MINDSET

Being an entrepreneur means embracing the fact that you're a lifelong student. The world is constantly evolving, and so should you. The strategies that worked last year might be obsolete today, and the tools you rely on now could be outdated tomorrow. This is why the most successful business leaders make it a point to educate themselves continuously. They understand that learning isn't just about gaining new skills—it's about staying relevant in an ever-changing market.

Think of it like this: if you're not learning, your competitors are, and if they're learning, they're gaining an edge over you. A study by the Harvard Business Review found that companies that invest in continuous learning are 46% more likely to be market leaders. That's not just a statistic—that's a wake-up call.

2. MAKE TIME FOR LEARNING

One of the biggest challenges entrepreneurs face is finding the time to learn. When you're juggling a million tasks—running a business, managing a team, dealing with clients—it can feel like there's no room left for anything else. However, here's the thing: if you don't make time for learning, you're doing your business a disservice.

The good news is that continuous education doesn't have to be a time-consuming ordeal. It's about incorporating learning into your daily routine in small, manageable ways. Set aside 15-30 minutes a day to read industry articles, watch a webinar, or listen to a podcast. These bite-sized learning sessions can add up, keeping you informed and inspired without overwhelming your schedule.

3. DIVERSIFY YOUR LEARNING SOURCES

Learning doesn't have to be confined to textbooks or formal education. In fact, some of the best knowledge comes from diverse sources. Read books from different genres, watch documentaries, attend industry conferences, take online courses, and—most importantly—learn from the people around you.

Your team, your clients, and even your competitors can be valuable sources of insight. Surround yourself with people who challenge you, who bring new perspectives, and who can teach you something you didn't know before. Don't be afraid to ask questions—sometimes, the most valuable lessons come from the simplest conversations.

4. KEEP UP WITH INDUSTRY TRENDS

In any business, staying ahead of industry trends is crucial. The market is constantly shifting, and what worked yesterday might not work tomorrow. Whether it's advancements in technology, changes in consumer behavior, or new regulations, being aware of these shifts can help you adapt and stay competitive.

Subscribe to industry newsletters, join professional networks, and follow thought leaders on social media. These resources can keep you in the loop and ensure that you're not caught off guard by sudden changes. Remember, being proactive about learning today can prevent you from scrambling to catch up tomorrow.

5. LEARN FROM FAILURE

Education isn't just about formal learning—it's also about learning from your experiences, especially your failures. Every mistake, every setback, every "oops" moment is an opportunity to learn something new. The key is to approach failure with a mindset of curiosity rather than defeat.

Ask yourself: What went wrong? What could I have done differently? How can I apply this lesson moving forward? By treating failures as learning experiences, you not only grow as an entrepreneur but also become more resilient. After all, some of the most valuable lessons are the ones learned the hard way.

6. DON'T NEGLECT GENERAL KNOWLEDGE

While staying informed about your industry is critical, don't overlook the importance of general knowledge. Understanding the world beyond your niche can provide valuable context and help you make more informed decisions. Whether it's learning about economics, history, psychology, or even the arts, broadening your horizons can give you a fresh perspective on business challenges.

For instance, understanding behavioral psychology can help you better understand your customers' motivations. Knowing a bit about global economics can inform your pricing strategies. Even learning about history can offer insights into how societal changes might impact your industry. The point is that the more you know, the better equipped you are to navigate the complex world of entrepreneurship.

7. EMBRACE THE UNKNOWN

Finally, remember that the pursuit of knowledge often leads you into uncharted territory. Embrace it. Don't shy away from learning something new just because it's outside your comfort zone. The world is full of unknowns, and as an entrepreneur, your ability to adapt and learn on the fly is one of your greatest assets.

Whether it's diving into a new technology, exploring a different market, or simply picking up a book on a topic you know nothing about, stepping outside your comfort zone is where real growth happens. Who knows, the knowledge you gain might just be the key to your next big breakthrough.

8. THE BOTTOM LINE: NEVER STOP LEARNING

Education doesn't end with graduation—it's a lifelong journey. The most successful entrepreneurs are those who remain curious, who continuously seek out new knowledge, and who aren't afraid to challenge themselves. So, make learning a priority, keep your mind open, and always be on the lookout for opportunities to grow. In the fast-paced world of business, the only way to stay ahead is to keep learning—and that's a lesson that never gets old.

LESSON 42: NEVER TEST THE DEPTH OF THE RIVER WITH BOTH FEET

We've all been there—a new opportunity presents itself, and you feel that rush of excitement. Your instincts tell you to jump in headfirst, to commit everything you have because this could be the big one. But here's the thing: jumping in without testing the waters first can lead to trouble. As the legendary investor Warren Buffett wisely said, "Never test the depth of the river with both feet."

1. THE TEMPTATION OF GOING ALL IN

It's easy to get caught up in the thrill of a new opportunity. Maybe it's a promising business venture, a new investment, or an innovative idea that feels like a sure thing. In these moments, it's tempting to throw all your resources into the mix, believing that the more you commit, the greater your reward will be.

Here's the reality: business is volatile, people are unpredictable, and the world never stops changing. No matter how confident you are, there's always an element of risk involved. That's why it's crucial to approach new opportunities with caution—by testing the waters before diving in.

2. THE WISDOM OF DIVERSIFICATION

The concept of not putting all your eggs in one basket is one that's been repeated time and time again, and for good reason. Diversification is one of the most effective strategies for managing risk. It's about spreading your resources across different ventures, investments, or strategies, so that if one fails, you're not left with nothing.

This principle doesn't just apply to financial investments; it applies to every aspect of business and life. Whether you're launching a new product, entering a new market, or even making a career change, it's important to have multiple options and safety nets in place.

3. BUSINESS IS UNPREDICTABLE

Let's face it—no matter how much research or planning you do, there's always a chance that things won't go as expected. Market conditions can change, customer preferences can shift, and unforeseen challenges can arise. When you go all in on a single plan, you leave yourself vulnerable to these uncertainties.

By keeping one foot on solid ground, you give yourself the flexibility to adapt. If the river turns out to be too deep, you can pull back and reassess without losing everything. This approach allows you to make informed decisions and adjust your strategy as needed rather than being locked into a single, potentially disastrous path.

4. THE BACKUP PLAN

Having a backup plan isn't a sign of weakness or lack of confidence —it's a sign of wisdom. A backup plan is your safety net and your insurance against the unexpected. It's what keeps you from drowning when the river turns out to be deeper than you thought.

For example, if you're considering investing in a new business venture, start by allocating a portion of your resources rather than everything. Test the market, gather feedback, and see how things unfold. If the results are promising, you can gradually increase your investment. However, if things don't go as planned, you'll still have resources available to explore other opportunities.

5. THE MYTH OF THE ALL-IN SUCCESS STORY

We've all heard the stories of entrepreneurs who went all in and hit the jackpot—the ones who bet everything on a single idea and came out on top. While these stories are inspiring, they're the exception, not the rule. For every success story, there are countless others who went all in and lost it all.

The truth is, playing it smart doesn't mean you're not committed —it means you're committed to succeeding in the long run. It's about balancing risk with caution and ambition with prudence. Warren Buffett, one of the most successful investors in history, didn't amass his fortune by taking reckless risks. He built it by making calculated decisions and always keeping a backup plan.

6. APPLYING THIS LESSON IN EVERYDAY LIFE

The wisdom of not testing the depth of the river with both feet applies to more than just business—it's a valuable lesson for life in general. When faced with any kind of risk, whether it's a financial decision, a career move, or even a personal relationship, it's important to protect yourself.

This doesn't mean you should be afraid to take risks. On the contrary, taking risks is essential for growth and success. However, it does mean you should take those risks with your eyes open and your feet firmly planted. Step one foot into the river to test the waters, and if it's too deep, you'll still have the other foot on dry land to keep you from falling in.

7. THE BOTTOM LINE: BALANCE AMBITION WITH CAUTION

In the world of entrepreneurship, ambition is key—but so is caution. While it's important to pursue new opportunities and take risks, it's equally important to protect yourself from the uncertainties that come with them. By following the wisdom of testing the waters before diving in, you can navigate the challenges of business with confidence and resilience.

So the next time you're faced with a promising opportunity, remember Warren Buffett's advice: never test the depth of the river with both feet. Keep one foot on solid ground, and give yourself the flexibility to adapt and succeed, no matter what the future holds.

LESSON 43: EXPERIENCE BEATS THEORY EVERY TIME

There's a saying attributed to Elon Musk that's bound to raise some eyebrows: "Drug dealers know more about running a business than 95% of college professors." Now, before you get caught up in the controversy, let's unpack the underlying truth in this bold statement. The essence here is simple: real-world experience trumps theoretical knowledge every single time. It's a reminder that business isn't just about what you learn in a classroom—it's about what you learn on the streets, in the trenches, and through the daily grind of running a business.

1. STREET SMART VS. CLASSROOM SMART

There's a clear distinction between being street smart and classroom smart. Classroom smart is all about theories, models, and hypothetical scenarios. It's what you get from textbooks, lectures, and academic discussions. But street smart? That's the knowledge you gain from real-world experience. It's the kind of learning that happens when you're out there making deals, solving problems, and navigating the unpredictable terrain of the business world.

Sure, a college professor can teach you the fundamentals of business—how to write a business plan, calculate financial projections, and analyze market trends. Yet, it's the business practitioner, even one operating in the gray areas, who understands the day-to-day realities of running a business. They know the hustle, the risk-taking, the need to adapt on the fly, and the importance of intuition—lessons that can't be fully captured in a classroom.

2. THE FEAR FACTOR

Starting something new, especially in business, is naturally accompanied by fear. Stepping into the unknown is scary. Yet here's the thing: fear is just an emotion, and like any emotion, it's something you can learn to control. What's not okay is letting that fear hold you back from taking action.

There's a reason why the saying "Fortune favors the bold" has stood the test of time. Success in business often requires taking risks, making bold moves, and venturing into uncharted territory. While fear can be a powerful deterrent, it's important to remember that it's just that—an emotion. It doesn't define you, and it certainly doesn't have to dictate your actions.

3. BUILDING YOUR BRAVERY MUSCLES

Overcoming fear isn't something that happens overnight. It's a muscle you have to build, and like any muscle, it gets stronger with practice. The more you step outside your comfort zone, the easier it becomes. Start by taking small risks—what I like to call "baby steps." Maybe it's launching a new product line, trying out a new marketing strategy, or making that cold call you've been putting off. Each step, no matter how small, is a victory in itself.

As you continue to push yourself, you'll start to notice that what once seemed terrifying now feels manageable, even exciting. You'll build resilience, gain confidence, and develop a mindset that welcomes challenges rather than shying away from them. This is the kind of learning that no textbook can provide. It's the kind of learning that only comes from doing.

4. THE VALUE OF REAL-WORLD EXPERIENCE

There's a reason why seasoned entrepreneurs often talk about the importance of getting your hands dirty. Real-world experience forces you to confront challenges head-on, make tough decisions, and adapt quickly. It's where you learn how to pivot when things don't go as planned, how to negotiate deals, and how to manage the unexpected twists and turns of running a business.

This isn't to say that formal education doesn't have its place—it absolutely does. Theories, models, and academic knowledge provide a valuable foundation. However, it's experience that turns that foundation into something actionable and practical. Experience is what bridges the gap between theory and reality, giving you the tools you need to succeed in the real world.

5. THE BOTTOM LINE: LEARN BY DOING

At the end of the day, the most valuable lessons in business come from doing, not just studying. Street smarts will always give you an edge over classroom smarts because they're grounded in reality. They come from facing challenges, making mistakes, and learning from those experiences.

So, if you're just starting out, remember this: don't let fear hold you back. Take those first steps, no matter how small they may seem. Build your bravery muscles, embrace the uncertainty, and learn by doing. Because in the world of business, experience beats theory every single time.

PART 7: SUSTAINABLE SUCCESS

LESSON 44: IF IT'S NOT BROKEN, FIX IT

There's a common saying that goes, "If it ain't broke, don't fix it." While this may apply to certain situations in the world of startups, I strongly disagree. In a startup, there's always room for improvement. Whether it's boosting productivity, using resources more efficiently, acquiring new skills, or simply making things better, you should always be proactive and plan a few steps ahead— even before problems arise.

1. START WITH WHAT'S PARTIALLY BROKEN

If something isn't working as well as it could be, don't wait for it to become a full-blown issue. For example, if you notice that your team structure isn't functioning at peak productivity—even if clients are receiving their work on time—don't settle for the status quo just because it's the easier decision.

In a marketing agency like OBCIDO, there are countless ways to structure the team. Each agency has its own approach to handling client communication, content creation, reporting, strategy setting, and more. Over the last four years, I've had to adjust our strategy no less than three times, often moving people around, creating new positions, or eliminating old ones. While these changes might initially frustrate your employees, it's crucial to implement them and explain that the goal is to boost productivity, enhance client success, and ultimately increase revenue. These improvements benefit not just the founder, but the entire team.

2. EMBRACE CONTINUOUS IMPROVEMENT

Even when things are going smoothly and your SOPs (Standard Operating Procedures) seem to be working fine, there's always room for improvement—especially with the rapid pace of technological advancement. New technologies like AI, business analytics, machine learning, blockchain, cloud computing, automation tools, and data visualization platforms are being developed every day, offering opportunities to streamline processes and achieve better results.

Take the time to explore these resources. Analyze your data, visualize it, communicate it effectively, and see how you can leverage it for better outcomes. For example, if you're closing three new clients each month, that might sound great on the surface, but if this closing rate is low compared to the number of sales calls you're making, it's time to dig deeper. Identify the underlying issues—whether it's your filtering process, expectation setting, negotiation skills, or follow-up strategy—and figure out how to improve.

3. THE IMPORTANCE OF PROACTIVITY

Being proactive means thinking ahead and addressing potential challenges before they become problems. This is a critical skill for any entrepreneur, and it's something you should instill in your team as well. One effective strategy is to empower your employees to take ownership of the challenges they face. Before they bring an issue to you or their direct manager, encourage them to write it down clearly, suggest two or three potential solutions, and then seek your guidance.

This approach not only fosters a culture of proactive problem-solving but also ensures that your team is constantly thinking about how to improve and move forward. Don't allow them to simply bring you problems without first considering solutions.

4. NEVER SETTLE FOR 'GOOD ENOUGH'

The lesson here is simple: don't settle for 'good enough.' Even if something isn't broken, there's almost always a way to make it better. By continuously seeking out opportunities for improvement, you're not just maintaining your business—you're actively pushing it to grow and evolve.

In the fast-paced world of startups, standing still is not an option. By adopting a mindset of continuous improvement and proactive problem-solving, you'll ensure that your business stays ahead of the curve and continues to thrive, no matter what challenges come your way.

LESSON 45: SUCCESS IS A MARATHON, NOT A SPRINT

You know how everyone wants things to happen fast? We live in a world where instant results are the norm—whether it's binge-watching an entire season of a show in a day or expecting a new business idea to take off overnight. Let's get real for a minute. Not everything works like that, especially in business. As Warren Buffett famously said, "You can't have a baby in one month by getting nine women pregnant." It's a funny way of saying that some things just take time, no matter how hard you push.

Success is one of those things. Sure, we all want it now, but the truth is, real success—the kind that lasts—comes from consistent effort over time. So, don't stress if things aren't moving as quickly as you'd like. The key is to stay focused, keep putting in the work, and remember that some of the best things in life take time.

1. THE LONG-TERM MINDSET

In today's world, where instant gratification is often the norm, it's easy to fall into the trap of expecting immediate results. Whether you're launching a new product, starting a business, or investing in the stock market, there's a natural desire to see quick returns. Understand this: true success rarely happens overnight. It's the result of consistent, long-term effort, not quick wins.

The most successful businesses and individuals are those who think long-term. They understand that while short-term gains can be satisfying, they are often fleeting. Instead, they focus on building something sustainable, something that will stand the test of time. This requires patience, persistence, and the ability to stay focused on the bigger picture, even when immediate results are lacking.

2. EMBRACE THE JOURNEY

One of the hardest lessons for any entrepreneur or investor to learn is that not everything will take off right away. You might launch a product that doesn't sell as well as you'd hoped or start a business that takes longer to gain traction than you anticipated. It's easy to get discouraged when things don't go according to plan, but this is all part of the journey.

Success is not a straight line—it's a winding road with ups and downs, detours, and unexpected challenges. The key is to stay committed, even when the results aren't immediately visible. Remember, every great success story is built on a foundation of consistent effort and resilience. Don't let early setbacks or slow progress shake your confidence. Instead, view them as opportunities to learn, adapt, and grow.

3. THE IMPORTANCE OF FEEDBACK

While it's important to be patient and think long-term, that doesn't mean you should ignore feedback. On the contrary, feedback is a crucial part of the process. It's how you learn what's working, what's not, and where you need to make adjustments. Here's the catch—feedback should be used as a tool for improvement, not as a reason to give up.

If your initial efforts don't yield the results you were hoping for, don't throw in the towel. Take the feedback you receive, analyze it, and use it to refine your approach. Success is often the result of many small adjustments made over time. It's the ability to iterate, pivot, and improve that ultimately leads to long-term success.

4. CONSISTENT EFFORT OVER TIME

One of the biggest misconceptions in business is that success comes from a single brilliant idea or a stroke of luck. In reality, success is the result of consistent effort applied over time. It's about showing up every day, putting in the work, and making incremental progress. It's the small, consistent actions that compound over time and lead to big results.

Think of it like planting a tree. You can't expect it to bear fruit the day after you plant it. It needs time to grow, to develop roots, and to mature. With the right care and attention, it will eventually produce a bountiful harvest. The same is true in business. Patience and persistence are your greatest allies.

5. AVOID THE TEMPTATION OF SHORTCUTS

In the quest for success, there's always the temptation to take shortcuts. Whether it's cutting corners, making compromises, or chasing quick wins, shortcuts can seem like an easy way to achieve your goals faster. In the long run, however, they often do more harm than good.

Shortcuts can lead to mistakes, damage your reputation, and ultimately set you back. True success is built on a solid foundation, one that can withstand the test of time. This foundation is created through careful planning, consistent effort, and a commitment to doing things the right way, even when it takes longer.

6. THE BOTTOM LINE: SUCCESS TAKES TIME

At the end of the day, success is a marathon, not a sprint. It's about playing the long game, staying committed to your goals, and understanding that some things just take time. Whether you're building a business, growing an investment, or pursuing a personal goal, remember that patience, persistence, and a long-term mindset are the keys to lasting success.

Don't get discouraged if things don't happen as quickly as you'd like. Keep putting in the work, stay focused on your vision, and trust that the results will come in due time. After all, as Warren Buffett reminds us, some things simply cannot be rushed.

LESSON 46: WINNING ATTITUDES ARE CONTAGIOUS

There's a lot to be said about the company you keep. You've probably heard the saying, "You are the average of the five people you spend the most time with." Well, there's a reason it's repeated so often—it's true. The people you surround yourself with have a significant impact on your mindset, your behavior, and ultimately, your success.

1. THE POWER OF POSITIVE INFLUENCE

It's better to hang out with people who are better than you. This might sound intimidating at first, but it's one of the smartest moves you can make. When you associate with people whose behavior, work ethic, and attitudes are superior to your own, you naturally start to elevate your own game. Their winning attitudes rub off on you, pushing you to strive for more, aim higher, and achieve greater things.

Think of it as a form of social osmosis. Just by being around high achievers, you absorb their habits, their mindset, and their drive. You start to see how they think, how they solve problems, and how they overcome challenges. Over time, their winning attitude becomes part of your own approach to life and business.

2. CHOOSE YOUR CIRCLE WISELY

One of the best ways to ensure you're surrounding yourself with the right people is by choosing your circle wisely. This doesn't mean you should cut ties with everyone who doesn't fit the mold of a high achiever, but it does mean being intentional about who you spend the most time with.

Seek out people who inspire you, who challenge you to be better, and who are living the kind of life or running the kind of business you aspire to. These are the people who will help you grow, who will hold you accountable, and who will push you out of your comfort zone. Whether it's through a professional network, a mastermind group, or simply a group of like-minded friends, make sure your circle is one that lifts you up rather than pulls you down.

3. THE VALUE OF A MASTERMIND GROUP

One of the most powerful ways to surround yourself with winning attitudes is by joining a mastermind group. A mastermind is a group of people who come together regularly to share ideas, offer support, and hold each other accountable. It's a space where you can brainstorm, problem-solve, and get honest feedback from others who are also striving for success.

The beauty of a mastermind group is that it's filled with people who are on the same journey as you, but each with their unique experiences, perspectives, and expertise. This diversity of thought can be incredibly valuable, giving you insights and ideas you might have never come up with on your own.

By joining a mastermind, you're not just surrounding yourself with high achievers—you're actively participating in a community that's dedicated to mutual growth and success. It's a place where winning attitudes are the norm and where you can't help but be inspired and motivated to reach your full potential.

4. THE RIPPLE EFFECT OF ATTITUDE

Winning attitudes are contagious, not just within your immediate circle but also beyond it. When you adopt a positive, success-oriented mindset, it influences everyone around you—your team, your clients, your family, and even your community. It creates a ripple effect, spreading positivity and motivation wherever you go.

As a leader, this is particularly important. Your attitude sets the tone for your entire organization. If you're optimistic, driven, and focused on solutions rather than problems, your team will follow suit. On the other hand, if you're negative, complacent, or constantly stressed, that attitude will also trickle down to those around you.

Remember, leadership isn't just about making decisions—it's about influencing others. One of the most powerful ways to influence others is through your attitude.

5. THE BOTTOM LINE: SURROUND YOURSELF WITH SUCCESS

If you want to succeed, surround yourself with people who are already succeeding. Winning attitudes are contagious, and by aligning yourself with high achievers, you set yourself up for success. Choose your associates wisely, seek out positive influences, and consider joining a mastermind group to keep yourself surrounded by the best.

In the end, it's not just about what you know or what you do—it's about who you surround yourself with. So, take a look at your circle. Are you surrounded by people who inspire and challenge you? If not, it might be time to make some changes. After all, success is a team sport, and the right team can make all the difference.

LESSON 47: ELIMINATE THE UNNECESSARY, ACHIEVE CLARITY

Bruce Lee, the legendary martial artist, once said, "It is not daily increase but daily decrease—hack away the unessential." This philosophy, rooted in simplicity and focus, is one of the most valuable lessons you can apply to both your personal life and your business. In a world full of distractions, noise, and endless options, the ability to eliminate the unnecessary is key to achieving clarity and success.

1. THE POWER OF SIMPLIFICATION

We often believe that adding more—whether it's features to a product, tasks to our to-do list, or complexity to our strategies—will make us more successful. The opposite is often true. The more you add, the more cluttered and complicated things become. Success, as Bruce Lee teaches, is about subtraction, not addition. It's about cutting out the nonessentials so that you can focus on what truly matters.

In business, this might mean streamlining your product offerings, simplifying your processes, or even narrowing down your target market. In life, it could mean decluttering your space, reducing distractions, or focusing on fewer, more meaningful goals. The goal is the same: to achieve clarity by removing the unnecessary.

2. AVOIDING THE TRAP OF "SHINY OBJECTS"

It's easy to get distracted by "shiny objects"—those new features, trends, or opportunities that seem so promising but can quickly pull you off course. While it's important to stay innovative and open to new ideas, it's equally important to stay focused on your core mission. Adding too many "extras" can dilute your brand, confuse your customers, and complicate your operations.

Think of your business or life as a straight line to your objective. The shortest distance between two points is a straight line, and every detour, distraction, or unnecessary addition only makes the journey longer and more complicated. By focusing on simplicity and eliminating the unnecessary, you stay on the most direct path to your goals.

3. CUTTING THROUGH THE COMPLEXITY

In today's fast-paced world, complexity is often seen as a sign of sophistication. Yet complexity can also be a trap. The more complex something is, the harder it is to understand, manage, and execute. Simplicity, on the other hand, brings clarity and efficiency.

To achieve this, you need to penetrate the complexity and go straight to the heart of the problem. What is the key factor that will drive your success? What are the essential elements that need your attention? Once you identify these, everything else becomes secondary. It's about cutting through the noise and focusing on what really matters.

4. THE ART OF SUBTRACTION IN BUSINESS

When you experience success, it's natural to want to build on it by adding more features, services, or products. However, sometimes, more isn't better. In fact, removing friction and complications from your offerings can lead to greater success and higher customer satisfaction.

For example, if you're running a product-based business, think about the user experience. Are there steps in the process that could be simplified? Are there features that aren't really adding value but are making the product harder to use? By focusing on what your customers really need and removing the rest, you create a cleaner, more effective product.

This principle applies to service-based businesses as well. Simplifying your service offerings, streamlining your processes, and cutting out unnecessary steps can make your business more efficient and your customers happier.

5. A SIMPLE LIFE IS A SUCCESSFUL LIFE

A successful life, much like a successful business, is often a simple one. It's about focusing on what truly matters and eliminating the rest. This doesn't mean your life or business should be bare-bones —far from it. It means that everything you do should have a purpose, and anything that doesn't should be let go.

Simplicity leads to clarity, and clarity leads to better decision-making, more focused action, and ultimately, greater success. It's about creating a life and business that is streamlined, purposeful, and free of unnecessary distractions.

6. THE BOTTOM LINE: LESS IS MORE

As you navigate the complexities of life and business, remember that less is often more. By eliminating the unnecessary, you create space for what really matters. Whether it's in your personal life, your business, or your products, simplicity is the key to clarity, efficiency, and success.

So take a step back, assess what's truly essential, and start hacking away at the rest. The path to success is often clearer—and shorter—when you travel light.

LESSON 48: CHECK YOUR EGO AT THE DOOR

Ego is a tricky thing. It's that voice in your head telling you you're the smartest person in the room, that you've got it all figured out, and that mistakes are for other people. It feels good, sure, but it's also dangerous. The truth is that ego can be one of the biggest obstacles to success.

We all have moments where we feel on top of the world—after a big win or a new milestone. It's easy for that pride to slip into overconfidence, making us believe we don't need anyone else's input. Here's the reality: no one knows everything, and thinking you do is a fast track to bad decisions.

The most successful people aren't those who think they have all the answers—they're the ones who keep learning, who aren't afraid to admit when they're wrong, and who understand that their ego can be their worst enemy. So before your ego takes the wheel, take a step back. Stay curious, stay humble, and remember that success isn't about always being right—it's about always being willing to grow.

1. THE EGO TRAP

Ego can be a powerful force. It can drive you to take risks, push boundaries, and strive for greatness. Nonetheless, it can also be your downfall. When you let your ego take control, you start to believe that you're infallible, that your ideas are bulletproof, and that you don't need to listen to anyone else. This is where the danger lies.

Ego blinds you to your own weaknesses and flaws. It convinces you that you're always the smartest person in the room, and that can lead to some pretty big mistakes. It's essential to remember that no matter how much success you've achieved, you're still capable of making errors—sometimes big ones.

2. THE POWER OF HUMILITY

The antidote to ego is humility. Humility doesn't mean downplaying your achievements or pretending you don't know what you're doing. It means recognizing that you're human, you have limitations, and you're always capable of learning and growing. Humility allows you to see the bigger picture, to listen to others, and to admit when you're wrong.

By checking your ego at the door, you open yourself up to new perspectives, new ideas, and new opportunities for growth. You become more adaptable, more resilient, and ultimately, more successful.

3. BEWARE OF COGNITIVE BIASES

As humans, we're all subject to cognitive biases—those mental shortcuts and errors in thinking that can lead us astray. Confirmation bias, for example, leads us to seek out information that confirms our existing beliefs while ignoring anything that contradicts them. The Dunning-Kruger effect makes us overestimate our own abilities, especially in areas where we lack expertise.

These biases can cloud our judgment, distort our decision-making, and reinforce our ego-driven belief that we're always right. The key is to be aware of these biases and actively work to counteract them. This might mean seeking out dissenting opinions, questioning your assumptions, or simply taking a step back and re-evaluating your decisions.

4. EMBRACE MISTAKES AS LEARNING OPPORTUNITIES

Mistakes are inevitable, but how you respond to them makes all the difference. When your ego is in charge, mistakes are something to be feared, denied, or blamed on others. However, when you approach mistakes with humility, they become valuable learning opportunities.

Every mistake is a chance to improve, to refine your approach, and to become better at what you do. Instead of letting your ego convince you that you can't afford to be wrong, embrace the fact that mistakes are a natural part of the learning process. The more willing you are to acknowledge and learn from your mistakes, the faster you'll grow.

5. THE BOTTOM LINE: KEEP YOUR EGO IN CHECK

In business, as in life, ego is both a motivator and a potential pitfall. It can drive you to achieve great things, but it can also blind you to your own shortcomings. The most successful entrepreneurs are those who recognize the dangers of ego and actively work to keep it in check.

By embracing humility, being aware of your cognitive biases, and viewing mistakes as opportunities for growth, you'll be better equipped to navigate the challenges of entrepreneurship. Remember, you're not as smart as you think you are—and that's not a bad thing. It means there's always room to learn, to improve, and to achieve even greater success.

So, check your ego at the door, stay open to learning, and never stop growing.

LESSON 49: LEARN, PRACTICE, SHARE

As an entrepreneur, one of your greatest assets is your ability to learn new skills. The world of business is constantly evolving, and the only way to stay ahead of the curve is by continually developing your abilities. Learning doesn't stop with you—it's just as important to share your knowledge with your team, helping them grow alongside you.

1. THE IMPORTANCE OF CONTINUOUS LEARNING

In the fast-paced world of entrepreneurship, standing still is the same as falling behind. New technologies, market trends, and business strategies are always emerging, and the most successful entrepreneurs are those who never stop learning. Whether it's mastering a new software tool, understanding the latest marketing trends, or improving your leadership skills, continuous learning keeps you sharp and adaptable.

Let's be clear: learning new skills takes effort. It's not enough to simply read about a new concept or watch a tutorial. You need to dive in, get your hands dirty, and practice until the skill becomes second nature. This requires time, dedication, and a willingness to push yourself out of your comfort zone.

2. THE PRACTICE MAKES PERFECT PRINCIPLE

Once you've started learning something new, the next step is to practice. It's not enough to know the theory; you need to put it into action. Practice allows you to refine your skills, learn from mistakes, and build confidence. It's through repetition and real-world application that knowledge turns into expertise.

Think about it like this: if you're learning to play an instrument, you can't just read the music and expect to perform flawlessly. You need to practice regularly, make mistakes, and gradually improve. The same goes for business skills. Whether it's negotiation, public speaking, or digital marketing, practice is what transforms knowledge into competence.

3. SHARING KNOWLEDGE WITH YOUR TEAM

Learning and practicing new skills is crucial, but don't keep that knowledge to yourself. Sharing what you've learned with your team not only helps them grow but also strengthens your business as a whole. When your team is equipped with the latest skills and knowledge, they're better prepared to tackle challenges, innovate, and drive the business forward.

Encourage a culture of learning within your team. Offer training sessions, share resources, and create opportunities for team members to develop new skills. When you invest in your team's growth, you're investing in the future of your business.

4. THE RIPPLE EFFECT OF KNOWLEDGE SHARING

When you share your knowledge, you create a ripple effect. Your team members, in turn, can pass on what they've learned to others, creating a culture of continuous improvement and learning. This not only enhances the capabilities of your team but also fosters a sense of collaboration and mutual support.

By sharing what you know, you also reinforce your own understanding of the subject. Teaching others forces you to clarify your thoughts, answer questions, and consider different perspectives. It's a win-win situation: your team grows, and so do you.

5. BALANCING LEARNING WITH EXECUTION

While it's essential to keep learning and developing new skills, it's equally important to balance that with execution. Don't get so caught up in acquiring new knowledge that you forget to apply what you've learned. The goal is to use your new skills to drive tangible results in your business.

Set aside dedicated time for learning, but make sure you're also putting those skills into practice. Whether it's implementing a new strategy, improving a process, or launching a new product, the ultimate aim is to translate learning into action.

6. THE BOTTOM LINE: NEVER STOP GROWING

As an entrepreneur, your journey is one of constant growth and evolution. By committing to learning, practicing, and sharing new skills, you not only enhance your own abilities but also empower your team to succeed. Remember, the most successful businesses are those that embrace change, adapt to new challenges, and continually strive for improvement.

So, keep learning, keep practicing, and don't hesitate to share what you've learned with others. In the ever-changing landscape of business, this is how you stay ahead—and how you bring others along with you.

LESSON 50: FINISH FIRST, PERFECT LATER

As entrepreneurs, we often strive for perfection. It's natural to want everything to be just right before we launch a product, service, or even a new idea. However, here's the truth: in the fast-paced world of business, done is better than perfect. The key to success isn't waiting until everything is flawless—it's about getting started, delivering value, and improving as you go.

1. THE PERFECTIONIST TRAP

Perfectionism can be a double-edged sword. On one hand, it drives you to deliver high-quality work and set high standards for your business. On the other hand, it can paralyze you, preventing you from launching anything until it meets your ideal vision. The problem with this approach is that it often leads to missed opportunities, delays, and frustration.

If you wait until your product or service is perfect, you might never launch at all. Or, by the time you do, the market may have moved on, leaving you behind. Remember, your customers aren't expecting perfection—they're looking for a solution to their problem. If you can provide that, even imperfectly, you're already adding value.

2. START WITH WHAT YOU HAVE

When my company first launched, we didn't have a perfect product or service. In fact, it was far from it. Yet, we went to market anyway, knowing that we could make improvements along the way. The revenue from those early sales allowed us to upgrade our offering, refine our processes, and eventually sell at a much higher price point.

This approach not only helped us get off the ground but also gave us valuable feedback from real customers. We learned what worked, what didn't, and where we needed to improve. If we had waited until everything was perfect, we might still be waiting to launch.

3. THE POWER OF ITERATION

Iteration is a powerful concept in business. It's about making continuous, incremental improvements over time. Instead of trying to achieve perfection from the start, focus on getting a version of your product or service out the door, then iterating based on feedback and real-world experience.

This process allows you to test your ideas, learn from your customers, and adapt to changing market conditions. It's a cycle of constant improvement that leads to a better end result than trying to perfect everything before launch.

4. DONE IS BETTER THAN PERFECT

"Done is better than perfect" is a mantra that every entrepreneur should embrace. It doesn't mean you should settle for mediocrity or cut corners—it means understanding that perfection is often an unattainable goal. By prioritizing action and progress over perfection, you'll achieve more and learn faster.

In business, speed and agility are often more important than getting everything exactly right. The faster you can bring your solution to market, the sooner you can start generating revenue, gathering feedback, and making improvements.

5. LEARNING FROM IMPERFECTION

Launching an imperfect product or service doesn't mean you're accepting failure—it means you're committing to a journey of growth and learning. Every mistake or flaw is an opportunity to learn something new and make your offering better.

Customers appreciate businesses that listen to their feedback and evolve over time. By engaging with your audience and showing that you're dedicated to improvement, you build trust and loyalty. Your customers become part of the process, and their input helps shape the future of your business.

6. THE BOTTOM LINE: PROGRESS OVER PERFECTION

In the entrepreneurial world, waiting for perfection is a luxury you can't afford. The most successful businesses are those that take action, deliver value, and iterate along the way. Don't let the pursuit of perfection hold you back from making progress.

Finish first, and perfect later. By focusing on delivering a functional, valuable product or service now, you can refine and enhance it over time. This approach not only accelerates your growth but also keeps you adaptable and responsive to the needs of your customers.

Remember, success isn't about getting everything perfect—it's about getting it done and continuously striving to make it better.

CONCLUSION: EMBRACING THE JOURNEY OF SUSTAINABLE SUCCESS

As we reach the end of this book, it's important to reflect on the overarching message: success in business is not about shortcuts or quick wins; it's about consistency, resilience, and the willingness to embrace the journey, no matter how long it takes.

Throughout these lessons, we've explored the lessons learned from years of navigating the complex world of entrepreneurship. From understanding the value of patience and persistence to recognizing the importance of adaptability and continuous learning, these principles have been the foundation of my own journey and the success of the businesses I've built.

One of the most critical takeaways is that there is no single path to success. Each entrepreneur's journey is unique and shaped by individual experiences, challenges, and opportunities. What remains constant, however, is the need for a strong foundation— one built on integrity, value-driven leadership, and a commitment to excellence.

Remember, success is a marathon, not a sprint. It requires thoughtful planning, but also the flexibility to pivot when needed.

It demands hard work, but also the wisdom to work smart. Most importantly, it calls for a deep understanding of your own goals, values, and the impact you want to make in the world.

As you move forward in your own entrepreneurial journey, I encourage you to take these lessons to heart. Use them as a guide, but also allow room for your own insights and experiences to shape your path. Don't be afraid to take risks, learn from failures, and celebrate the small wins along the way.

Success is not just about reaching the destination—it's about growing, evolving, and enjoying the process of getting there. So, embrace the journey, stay focused on your mission, and remember that with perseverance and a clear vision, you can achieve sustainable success that lasts.

Thank you for allowing me to share my experiences and insights with you. I hope this book has provided valuable lessons and inspiration as you continue on your path to building something truly remarkable.

ACKNOWLEDGMENTS

First and foremost, I want to express my deepest gratitude to my mother, whose unwavering support and encouragement have been the foundation of everything I do. Your belief in me has been my constant source of strength, and your motivation has fueled my journey to chase my dreams. Thank you for always standing by me.

A special thank you to Chaker Khazaal, my mentor, and greatest supporter. You have been an instrumental part of my journey, guiding me through the challenges and inspiring me to push beyond my limits. Your wisdom and generosity have shaped me both personally and professionally, and I am forever grateful for your continued belief in me.

To my incredible team: none of this would be possible without you. Your dedication, hard work, and commitment have made all the difference. Together, we've achieved great things, and I am proud to be leading such a strong, talented group of people. You are the backbone of our success, and I thank you all from the bottom of my heart.

Finally, I want to extend my heartfelt thanks to our amazing clients. It has been an honor and privilege to work with you, and your trust in us has allowed us to grow, innovate, and learn. We've shared an incredible journey together, and I am grateful for the opportunities you've given us to help shape your companies and our future.

Thank you all for being a part of this journey with me.